CONSCIOUS AGRICULTURE: ALIGNING WITH NATURE

THEODORE CARLAT

CONSCIOUS AGRICULTURE: ALIGNING WITH NATURE

THE FUTURE OF FOOD

INTRODUCING THE AUSTRALIAN
DEMETER BIODYNAMIC METHOD

Theodore Carlat
Conscious Agriculture: Aligning with Nature

First Edition 2024

Published by Spines
ISBN 979-8-89569-822-8

*1. Agriculture 2. Bio Dynamic Farming 3. Organic Farming 4. Nature 5. Nutrition 6. Nutrient
Dense Foods 7. Regenerative Agriculture 8. Ecological Farming 9. Biological Organic Farming,
10. Sustainable Agriculture 11. Carbon Sequestration 12. Climate Change 13. Food Security 14.
Food History and Future 15. Soil Health and Fertility 16. Ecology 17. Soil Science*

To Ayla, my daughter who has embraced in her life work the light and sound, the twin pillars of consciousness. And, for her patience and intuition which our youth need to embrace for a more conscious future. To all young people who seek to support a living healthy planet and the pure nutritious healthy food they deserve.

CONTENTS

PREFACE

A conscious approach to agriculture is neither a new phenomenon nor a fad. The culture of Agri developed between eight to ten thousand years BCE in the fertile crescent of the Near East. Agriculture began with the Holocene, the age of man, or the Neolith Revolution. *The development of agriculture made civilization possible.* Beyond providing daily food, it made possible the bounty of scalable food production and surplus. Our modern-day imagination can only begin to comprehend the impact these activities continue to have on our lives today.

Never have the challenges facing agriculture—and its impact on humanity—been greater than they are now. Soil is undergoing an extreme process of degradation worldwide, resulting in massive loss of topsoil and its loss of biology. With seventy percent of freshwater being used for agricultural purposes, global freshwater supplies are rapidly diminishing. Our current, longstanding agriculture practices have resulted in habitat destruction and a loss in habitat diversity—an ongoing, sixth mass global extinction. Insect populations and wild species have on average declined globally by forty percent or more. They are causing deforestation, desertification, and the decline or collapse of formerly abundant oceans. We are perpetuating a dramatic increase in atmospheric gases, fueling global warming, climate changes, and extreme weather. In the face of food insecurity and famine, the human population is growing while health and longevity are declining. In short, as a global society, we are on the edge of the ecological collapse of numerous environments and their species. Within the "industry of agriculture", an ongoing lack of conscious decision-making, destructive behaviors, and inexcusable inaction are rendering our planet direly ill. Solving the planetary crisis is not only necessary but also possible! We will all win if we succeed at restoring Nature.

And yet—all the solutions to meet today's vast agricultural and environmental challenges are available and viable if we can only put them into action. It will come down to this question: as a society, are we capable of changing our behaviors and patterns of consumption? And how will we respond to the state in which we find ourselves and our blue planet? Today it is cheaper to produce energy from renewable sources than from fossil-based energy. What choices will we make? Choices that continue to be driven by wants, desires, and self-gratification, or those that begin to rationally and consciously strive toward planetary well-being, harmony with nature, and self-preservation? It is our choice we are facing today! Time is running short. Necessary changes must happen in this generation, or we will face irreversible losses and decline for generations to come. As Neil de Grasse Tyson said, on the idea of the world ending: "We might have 100 years left!" Is this meant to be an asteroid extinction event? Or an extinction behavior of civilization by our own hands? The *Age of Nature* documentary series positions the role of agriculture as central to this tipping point:

> "How we live with nature now will determine our future...At this point, it's not enough to just produce food in ways that minimize harm to the planet—we must start producing food in ways that actively restore the health of the planet."

So, allow me to introduce you to the concept of *conscious agriculture*. Conscious agriculture involves the world's entire food ecosystem and society—from the farmer to the grower, to the environmentalists, to the chefs and foodies and everyday consumer—becoming a fully conscious food society will have a dual awareness of both our present state and future consequences of food production practices. Embracing conscious agriculture, in all its aspects, is the most important act we as human beings can undertake to bring our planet's health back into balance.

A study of conscious agriculture involves a deep understanding of the Australian Demeter Bio-dynamic Method—certainly the most widely used biodynamic system in Australia, and perhaps worldwide. In my research, of all the positive and effective biodynamic, regenerative, and biological-organic methods in use today, I have found the Australian Demeter method to be the most conscious agricultural practice. And why is that? The simplest and most needed solution to widescale, agriculture-driven challenges is creating good, productive, living agricultural soil. Specifically, it has been widely suggested that regenerative or soil-building agriculture utilizing composting methods holds the solution to sequestering much of the excess global atmospheric carbon, leading to increased plant growth and the restoration of grasslands, forests, estuaries, and tidal ecosystems. Put simply, reducing excess atmospheric carbon, nitrous oxide and methane from fossil fuels is the number one environmental priority for the planet today, and sequestering carbon and nutrient recycling in our soils is our future solution to food production. This is where all conscious thinking societies need to pay their full attention if we are to slow down and avoid a very costly continuation of extreme weather calamities presently plaguing our planet.

The earliest *form* of agriculture was animal husbandry—shepherding of domestic livestock, both goats and sheep, this was followed by pigs and cows. Humane care and the appropriate integration of animals—that is, good animal husbandry—was foundational to the first waves of subsistence and production agriculture systems. Breakthroughs began to be made by benefits from livestock dung or manure and well-managed grazing practices provided the basis for the soil's increased fertility; when livestock herds were properly managed, they resupplied and recycled the essential plant macro-nutrients NPK, necessary for agriculture to thrive.

For a long time, traditional agriculture was always a natural, "organic" practice. It didn't involve any synthetic materials (substances or inputs derived from Petrochemicals and concentrated nitrogen compounds) and little if any artificially imported fertility materials or minerals. In short, agriculture succeeded for centuries practicing simple, minimalist, or low-till soil cultivation, mostly recycling and regenerating plant nutrients.

And then came the Industrial Revolution, which ushered in practices of industrial agriculture—and everything changed. Agriculture now affects everything, every living thing and organism in the biosphere.

Just over two hundred years ago, along with widescale societal changes caused by the Industrial Revolution, we began burning exponentially more and more fossil fuels—coal, petroleum, and natural gas. At the same time, the advent of "chemistry-based agriculture" began the accelerated practice of death in soils and our environment. Since the Industrial Revolution began, excess atmospheric carbon has doubled.

These trends of decline accelerated exponentially from there, in line with major world events. In the 1950s and 60s, after WWII, growers began using water-soluble nitrogen fertilizer on a grand scale (ironically referred to as the Green Revolution). Toxic poisonous chemical sprays (insecticides, fungicides, and herbicides) followed, a relatively new threat to life. Today, millions of metric tons of deadly chemical cocktails are used annually in the production of most of the world's food supplies completely unnecessary.

When will these extremely deadly chemical practices cease? And how will Nature and the Earth recover? Our ancestors relied one hundred percent on sunlight as our fuel and food source. Today half or fifty percent of our proteins and much more of our calories from carbohydrates rely on fossil fuels to grow, produce, and sustain modern humans! We must become environmentally conscious very soon as a species to reverse the destructive trends we humans have unleashed. We must embrace conscious agriculture on a global scale, now! Our planet, life, and human civilization depend on these actions!

1

———

AGRICULTURE: WORKING WITH NATURE

Agriculture is our human species' ancient and sacred heritage. Agriculture is a bedrock stone of our entire civilization, of our entire history of human achievement. No matter how much we evolve and change, we cannot escape that our practice of agriculture—and all its consequences—will always be subject to nature and its natural laws. In modern society, we seem to have forgotten this. At one point in the not-too-distant past, agriculture was truly a way of life—a culture worth preserving and maintaining—and we must get back there. The stakes are too high. Participants in the global agricultural process must remember to work with nature, this is imperative!

We'll start at the beginning. When did agriculture begin? Where did it begin?

Our planet as a natural resource far predates any agricultural activity. Before the advent of agriculture—or, at the very beginnings of animal husbandry practice—hunter-gatherers at the beginning of the Holocene age of man engaged in food rituals involving astronomy and observation of the heavens. (To be sure, concepts of food and spirituality have been inextricably linked from the beginning.) More and more, ongoing archaeological discoveries and their evolving interpretations continue to challenge long-held beliefs that man learned the first rudiments of agriculture before practicing religion. Many now believe that agriculture grew from established practices of spirituality.

In 1882, *Oahspe*—a "New Bible"—claimed that man-built oracle houses and star observatories, such as *Göbekli Tepe*, Stone Henge, Newgrange, Carnac, and numerous ancient stone sites of worship came first and predated the advent of farming. Were these structures and chambers built so that God could instruct man, through seers, prophets, and sages in the knowledge of the seasons, so that man might learn the ways and wisdom of agriculture? Sumerian mythology purports that "sixty wisdoms" were given to mankind after the Great Flood by the Anunnaki gods, directly allowing for the development of associated industries and trades such as agriculture, arts, and crafts that supported settlement and thus civilization. Is it possible that humans were divinely inspired or instructed to take up agricultural practices?

This foundational relationship between agriculture and civilization continues to be just as relevant today—perhaps even more today than ever. How can we even begin to understand the modern dilemma around food, agriculture, the environment, and our climate? How can we truly grasp our relationship to nature and the food it provides us? As Rudolph Steiner put it: "We must do so in studying that which alone makes possible the physical life of man on Earth—and that is, after all, Agriculture." Or, as Alan Savory said: "Agriculture is the key to everything. It is the foundation upon which all human endeavor rests, but we treat it as simply dirt."

How could the answer be any clearer? The future and fate of human civilization as we know rests on the shoulders of agriculture practices and choices.

Then vs. Now

From Media (later known as Persia), Zarathustra (Zoroaster in Greek) was the father of Indo-European agriculture. He gave us foundational knowledge of the seven creations of life: fire, air, water, earth, plants, and animals, and the seventh creation, humans, meant to care for and elevate the natural world and its creations. This was some of the earliest wisdom given to humankind. Modern people and our civilizations have failed to uphold our responsibility and promise.

In many ways, agriculture can be understood as the first and longest continual culture developed by humanity. The culture of Agri began near the province and capital of Agri, located today in eastern Turkey, near Armenia and Mount Ararat. From Anatolia, the world's first culture of Indo-Europeans, the culture of "Agri" spread out over the millennia from the fertile crescent throughout the arable lands of the globe with the advent of animal husbandry, herding, and care of domesticated livestock. First came the relative domestication of livestock (sheep, goats, hogs, and cattle); this was followed by the cultivation of early ancient cereal grains (einkorn, spelt, emmer, rye, and barley); tilling of the soil (loosening, or disturbing, and opening the soil for the planting of cereal grain seed) began nine to ten thousand years ago. These developments seem to have been parallel with the development of domesticated rice in Asia, and corn (maize) in North and Central America. The first olive orchards were planted in Palestine eight thousand years ago, which then spread throughout the Mediterranean cultures.

Hunters and gathers were dependent on the wild natural world for food sources, a practice that sustained much smaller human populations. From the time of human origin, food has been difficult to obtain but nature is very bountiful and resilient. From the very first-world civilizations of the Indus Valley and Sumer, humankind has been dependent on the bounty of agriculture. Agriculture fundamentally changed the trajectory of survival for human populations and the development of civilization. (It has even been suggested that agricultural expansion perhaps influenced the actual evolution of human DNA, via widespread consumption of grains and dairy products.)

At times, in centuries past, agricultural surpluses were produced for their cultivators and their communities. Especially with the introduction of mixed livestock and diversified cropping, perhaps agriculture reached an apex as traditional old-world organic farming in Europe around five hundred years ago. But the earliest agricultural practices were all about survival. Resources were local and were found on hand. Little or nothing was imported onto the farmlands. Domestic animal manures and plant residue were utilized, recycling nutrients carbon, nitrogen, phosphorous, potassium, and trace elements to develop and improve soils and fertilize soil thus crops.

The wisdom of native and indigenous science practiced throughout the Americas, Central Asia, and Western Europe involves an understanding of the wholeness and innate intelligence of the natural world—and the importance of working with Nature. Essentially, this refers to an understanding of our larger, human connectedness with the stars, the Earth, and all living things. Today, tragically, this understanding has been largely lost.

Breaking terribly with our great knowledge inherited, modern cultures have misused (degenerated and degraded) the element of fire to foul the element of air, with excessive clearing and burning of natural landscapes (clear-cutting and burning of forests) and burning of fossil fuels for industry, electricity, and transportation. Nuclear energy pollutes and defiles the earth, air, and water. Disgraceful livestock practices, pollution of water systems and wells on mass, inhumane treatment of farm animals, and abusive livestock conditions violate an inherited sense of natural trust between living beings. The development and use of poisonous chemicals in growing our food—and the tragic advent of genetically (engineered) modified organisms (GMO crops)—has unleashed disaster onto the natural world. And it is often scientists—those who say they understand nature best—who pave the way for these tragic developments, through mostly materialistic practices. Many scientists consider life as a nonliving reality, as nothing more than a handful of random chromosomes that can be manipulated and destroyed without thought of life's sacredness; they often see nature as nothing more than a factory that can be driven to produce endlessly, without consequence.

Modern industrial agriculture is primarily driven by goals of production—that is, producing weight and volume. Today, most of what the agricultural machine produces is tainted with toxic chemicals aimed at producing massive quantities of commodities. "Foodstuffs" that, for the most part, are crops with empty calories, amounts to an *empty harvest*: foods lacking essential macro and micro-nutrients. Because this is what the machine of industrial agriculture is producing on an overwhelming scale, modern crops are incapable of providing the good, healthful nutrition that is required to sustain animal and human health and life in the long term. And their disastrous consequences don't end there. The broken nutrient cycle in our agricultural practices has disrupted healthy human states of nutrition and fertility. A multi-billion-dollar supplement and nutrition industry has developed as a Band-Aid, as foods chemistry-based agricultural foods are not adequate for providing vital nutrition and health. Of course, the original purpose of agriculture is and has always been to produce nutritious food and fiber that can clothe, feed, sustain, and grow the populations of society. In this, modern agriculture is failing in its ordained goal. The materialistic greed mentality is only about growing more and more volume!

What Is Driving the Downfall?

When you consider our current predicament holistically, the ironies and questions appear endless. It seems like a reasonable assumption that the defining end goal of agriculture would be, first and most importantly, growing nutritious food. Why is this not the driving objective in agricultural practice today? Is growing nutritious, wholesome food that will maintain and sustain our mental and physical well-being not essential? Aren't flavor and nutritional values at least as important as production volume? Mustn't we weigh the value of human health, soil, environmental health, and food quality over the value of volume production? Is it not important that plant and animal foods are vital, alive, and able to carry genetic diversity into the next generation?

So much of this comes down to unnatural interventions. Biotechnology is, by definition always attempting to alter the natural order.

Bioengineered food is anti-evolution and anti-nature itself. (Consider how unappetizing that sounds.) The results of these kinds of interventions are not always unnatural—just the methods. Take, for example, the ability of a plant to become more efficient at producing biomass, at performing photosynthesis. It is little known by industrial agriculture researchers that 501 the biodynamic horn silica material (from quartz crystal) is a natural material that assists plants' ability to better utilize light in photosynthesis. This solution already exists in nature's toolbox, but big, industrial agriculture is looking for an unnatural alternative. Corporate-driven research is not looking for simple or open solutions unless corporate patents can be held and greater profits can be made by agrochemical corporations.

Sometimes, it seems like every high-order industry has turned its back on the natural health of the planet. Agriculture has become "agri-business." Outdated scientific theories and corporate greed promote the use of poison fertilizers and pesticides, and the non-sustainable importation of chemistry inputs which degrade our soils, waterways, groundwater, and oceans deplete non-renewable minerals, and destroy soil carbon reserves. Why do the *academic experts*, with little or no practical farming experience, advise our farmers? Is agriculture to continue to be reduced to an industrial process? Why have we allowed industrial farms to basically, become "food factories"? To be clear, I am no romantic. Some people may dream and wish, at times, of a simpler world—the world of the hunter-gatherer, of the horticulturist, tending nature's bounty in the Garden of Eden —but these times are no more. For at least eighty-five percent of humanity, food is grown from the cultivation of crops or agriculture. Our mega-cities with populations in the millions and tens of millions are dependent on agriculture to provide food to their masses now more than ever. But if we don't change how we produce food for humanity—even on a mass, global scale—we are all doomed.

It All Comes Down to Soil

As a species, we rely primarily on good, healthy, fertile soil and its productivity for our physical sustenance. This is nature's law and system plants have evolved with for millions of years.

Working with nature starts with understanding the processes that nature uses to create soils. Soil is built from the top down. Opening the soil—tilling or cultivating the soil—is essential to good soil management and agriculture. More specifically, the *method* used to open the soil has the potential to activate half of the soil's needed fertility for a season's crops—when the right implements are used, and good timing is followed. Some voices advocate no-till agriculture, but this is neither realistic nor necessary. Cultivation brings carbon, nitrogen, light, warmth, and air into the earth. Good soil cultivation can improve the soil's condition, structure, and tilth, encouraging humus formation. Proper, low-till soil cultivation can release nutrients, stimulate soil organisms, and incorporate organic matter into the topsoil when done properly.

The "plow" associated with the inception of agriculture was not a plow at all, but the Ard or "scratch plow"—a wooden soil ripper. This cultivating tool was used for thousands of years to cultivate the earth. (Today, a few isolated traditional cultures still use the Ard and draft animals in Euro Asia.) Soil cultivation—that which does not radically destroy soil life—is at the heart of conscious agriculture.

Other methods stimulate and feed soil microorganisms too: the rotation of livestock on pastures and meadows; seasonal cropped plantings; manuring (composting); and use of cover crops. These are not new practices. Farming systems using livestock rotations, and "open field" cropping allowed human populations to double during the Medieval agricultural revolution (1,000–1,300, CE) due to increased farm productivity. During this same period, the introduction of the moldboard plow, which lifts and lays the soil on its side layer after layer without greatly destroying soil structure and soil life —was another important factor in the scaling of successful agriculture practices.

Today, commonplace plowing practices, techniques, and technology are excessive and extreme. The indiscriminate misuse of mechanical cultivation dramatically harms soil health. Large-scale industrial farming has devastated much of the arable chernozem soils worldwide. The practice of injecting ammonia nitrates as nitrogen fertilizers is destroying soil carbon. At a much higher level, massive losses of soil carbon stores, soil erosion, and soil compaction have been the result of modern conventional industrial agricultural machinery contributing to increased atmospheric carbon, now at a rate of 420-440 ppm.

Soil humus, a stabilized, complex, carbon-based material, was discounted and taken for granted in the early 1800s when the theory of chemistry-based agriculture was first introduced. Today it is estimated that 30% of the world's agricultural soil—1.7 billion hectares (over 4 billion acres) of topsoil—has been lost. Globally, the agricultural land area in use is *approximately five billion hectares,* about 38% of the global land surface. Cropland takes up 18% of the total acres of agricultural land. The remaining 20% consists of grassland meadows and pastures for grazing livestock.

It is the humus that determines the profitability of farming. Just a 1% increase in soil humus can dramatically increase the water-holding capacity of the soil, on one acre by 68,000 liters. Good soil, as defined by the USDA, has 4% organic humus content and one million earthworms per acre. In addition, healthy soil has 50% air and water spaces, good tilth, and no compaction. On the flip side, enormous topsoil losses represent carbon loss and humus loss—losses that have devastating consequences. They are threatening food security, increasing global weather extremes, rainwater runoff, and increasing extreme flooding events. Active "humus farming" and carbon sequestration should now be an essential global concern and the number one priority of every nation. Soil biology and nature's system of fertility must become widespread common knowledge.

For over one hundred years, "agricultural chemistry" has been responsible for the mass degradation and the wholesale destruction of humus soils. Chemically managed/treated and heavily cultivated soils may have as few as 70 to 80 thousand earthworms per acre—far short of 1,000,000 per acre in healthy soil. Farm soils below 2 to 2.5% humus (organic matter) are not agriculturally productive at all. Most terrifyingly, via the industrial farming system, billions of tons of sequestered soil carbon have been released worldwide. It is estimated that half of the excess atmospheric carbon accumulation in the last two hundred years has been caused by the destruction of soils.

Against this backdrop, the human population has reached eight billion, quadrupling since 1900. At the end of the last Ice Age, in the face of glacial melting and global flooding, as few as 10,000-100,000 humans are thought to have seeded the modern human race. At the time of the dawn of agriculture—around 8,000 BCE, the human population of the world was approximately 5 million. It grew over the next 8,000 years to perhaps 200 million in 1 CE (some estimate 300 million or even 600 million, suggesting how imprecise population estimates of early historical periods can be), with a growth rate of under 0.05% per year. Essentially, the point is this: the human population was originally somewhat sustainable. Today the opposite is the case: human populations may not be sustainable, and growth rates could very likely outpace agricultural production and resources if left unchecked. Is this a myth or is our mismanagement of agricultural land, misdirected modes of production, and inequitable distribution of the food produced the real problem?

What does all of this mean for our future survival?

Moving Toward a Biodynamic Renaissance

For more than ten thousand years, agriculture was biological and organic. Chemistry-based agriculture did not exist at all. Nothing artificial synthetic was used in the ancient world to grow food. Nothing was used in original agriculture practice that was unnatural or did not come from living nature. Until the industrial revolution, the importation to the farm of outside fertility nutrients—farm "inputs"—was very rare. The few historical exceptions to these trends, such as early cases of applying nitrogen and phosphorous-rich bird or bat guano and mineral soil amendments, were highly rare, until about two hundred years ago.

Reductionist-materialist concepts of science posit that when plant nutrients are exported off a farm, with farm products, new plant nutrients must be imported back into the farm to replace what has been lost. This would mean that, when 100% of a farm's harvest is sold each season and shipped away, all nutrients must be replaced. This materialistic "first theory of agriculture" may seem reasonable, but it represents a shortsighted understanding of natural science.

Globally, millions of megatons of plant nutrients; powdered rock, and water-soluble nitrogen are added to agricultural soils each year. A pound of lime (calcium) or phosphorus (rock phosphate) added to a farm's soil will not return a pound of that mineral to a farmer's grown crops nor to the food produced from that crop. Minerals applied may have as little as 1-2% utilization by crops and long-term plants may have much less than 1% use of these nutrients—all paid for by the farmers and then the consumer. These are materialistic strategies unsupported by scientific facts. So much is wasted or lost in the subsoil and waterways in these processes that it looks like a marketing scheme more than good science. This is a great environmental and economic waste, to lose these valuable farm nutrients. Fifty percent loss of applied water-soluble nitrogen is common and even eighty-five percent losses are acceptable.

How is this good science? These theories do not account for the natural activity of soil biology and good farm management. In a balanced ecological farming system, nature can and will regenerate the soil's nutrients its minerals. Active soil biology and recycling some of the farm's nutrients in a closed system or a farm organism can regenerate the soil. A holistically functioning agricultural system limits the quantities of products exported, perhaps to sixty percent maximum. Sadly, advocates of an industrial system of agriculture justify their techniques by the large yields of foodstuffs they believe can only be produced and harvested using artificial interventions. Some truly believe that humanity today can't survive without the large-scale, wasteful, highly damaging applications of phosphorous, potassium, and synthetic, water-soluble nitrogen, so they simply accept the evils of industrial agriculture—waste of nitrogen and phosphorous, pollution to the environment, toxicity in produce, increased costs for farmers, diseases caused by water-filled crops and pesticides, et cetera—as necessary.

The good news is this – all hope is not lost. Today, innovations from a creative new generation of farmers – drawing inspiration from the simple, inherited wisdom of peasant farmers in generations past and new insights offered by Dr. Rudolph Steiner – hold the future of food. Old-world peasant farmers grew their food organically, with great attention and care, and this approach is having a modern revival, being blended with new thinking and science in the realms of agroecology, agroforestry, and biodynamic—regenerative-*organic*— agriculture. In short, a new renaissance in agriculture is at hand, with the emerging knowledge of biodynamics, soil biology, and soil humus-building strategies being implemented alongside traditional knowledge We need biological organic, not more chemistry-based agriculture.

So, what is biodynamic agriculture? Biodynamic agriculture is a truly self-sustaining system of agriculture, one that supports soil to meet its own fertility needs without importing outside fertility additives. This is accomplished once soil biology is stabilized and a productive fertility level is built up. When soils are in harmony once again, there is no need for inputs or additives: soil health and fertility and a closed nutrient cycle are achieved, farm health of both plants and livestock are maintained, and productivity results.

In the early years of the last century, new agricultural philosophy and science emerged from the vision and insights of Dr. Rudolph Steiner, whose ideas gave rise to the seeds of a truly holistic, organic, regenerative agricultural system—ideas that have since developed into a practical, worldwide regenerative-organic "biodynamic" approach to agriculture inspiring the beginnings of most of the world's organic movements over the last eighty years.

In 1923, Dr. Rudolph Steiner was asked to address declining health and fertility issues emerging in agriculture at that time. In

1924, Steiner gave a series of eight lectures that became the foundation of the biodynamic method known as *The Agriculture Course*, the first distinct form of organic agriculture. Perhaps what made his perspective so memorable and impactful was his blending of the scientific and the spiritual harkening back to agriculture's very origins on earth. (Spiritual here refers to the unseen forces or influences behind physical phenomena and the intelligence-driven reality all around us; and does not refer to the random, accidental existence often attributed to life by an atheistic-driven scientific worldview).

Biodynamic agriculture was an organized system in America by 1938. The term biodynamics was first coined in 1938 in Ehrenfried Pfeiffer's book *"Bio Dynamic Farming and Gardening."* New Zealand, Poland, Germany, Switzerland, and Australia had biodynamic farming practiced by the mid-late 1920s. In 1940 in England, Albert Howard published his book *An Agricultural Testament.* He inspired the organic movement after studying traditional agricultural methods in India and collaborating with early biodynamic pioneers. Australia's biodynamic research began early also and developed in the mid-1950s and early 1960s. In America, the organic movement finally began to develop in the 1960s and 70s, largely spurred by rising public awareness of synthetic substances (pesticides) and their chemical agricultural abuses due to the bestselling book *Silent Spring* (Racheal Carson, 1962) and Rodale's *Organic Gardening Magazine.*

Biodynamic agriculture is needed now more than ever, as a new impulse in soil biology and life-promoting agriculture. Thanks to the ever-expanding knowledge of the soil food web, we now have a great understanding of a myriad of available biodynamic solutions: the practices of composting, the use of compost teas, proper aerating of the soil, diverse plantings, and the return to the holistic use of mixed livestock providing fertility, to name a few. Also—after being marginalized for decades based on advice from modern scientific-chemical agriculture "experts"—many traditional biology-driven organic methods are returning and being practiced under many names.

In short— biological agriculture is having a great renewal. Consumers are demanding un-poisoned and nutritious food.

However, policymakers in government and the scientific community have shown little or no recognition or commitment to the 100% complete conversion to organic agriculture, as reflected by the lack of funding, few academic studies, and limited scientific objectivity over the past several decades. While the <u>US Department of Agriculture (USDA)</u> has had sincere leaders heading their <u>Certified Organic Program</u>—the program that certifies organic food and is aligned with international standards— the USDA is headed by the corporate puppeteers of industrial agriculture "Monsanto" serving their corporate masters. This political-corporate marriage is a great disservice to the average American citizen, that perpetrates a prejudice against "organic food." (Corporate interests are against even labeling GMO foods in the United States.) Unbiased research and news releases about "organic food" have been comparatively few. Relevant university studies of organic or biodynamic agriculture and food conducted over the last few decades have been pathetically lacking.

The good news is that, finally, some organizations are emerging to fight back. The <u>International Federation of Organic Agricultural Movements (IFOAM), or Organics International (founded in 1972)</u>, is the worldwide umbrella organization for the organic agriculture movement, which represents close to 800 affiliates in 117 countries. They seek to be a part of the global solution, assisting farmers to transition to organic agriculture, educating on and promoting the consumption of organics, and advocating on a policy level for agroecological farming practices and sustainable development.

In the US it is reported that there are 5.68 million acres or 2.03 million hectares of certified organic land. We have an opportunity here at home to do much more, both for our well-being and that of our global community. The US ranks only fifth in the world of organic agriculture with 1.2% of acreage in organic certification, after Argentina, China, Spain, Germany and Australia. Globally, Australia is leading the way. With 35.64 million hectares of land producing 54% of global certified organic-biodynamic foods, more than 50% of all certified organic farming in the world today is practiced in Australia. 8.8% of all developed agricultural lands in Australia are organic and biodynamic, with improved grazing pastures and cultivated agricultural lands, wild grazing vegetation not included—making it one of the first countries and continents to achieve this milestone and eco-victory competing alongside industrial chemical agriculture. A substantial volume of this organic farming in Australia is Demeter-certified biodynamic farming. Demeter is the umbrella organization that certifies and promotes biodynamic agriculture internationally in over sixty-two countries.

Presently, we need to move from a reawakening to a revolution in public awareness, with the world demanding biological-regenerative organic food. Our collective goals must be: (1) producing enough quality, nutritious food, (2) creating and supporting healthy ecosystems with "on-farm fertility", (3) healthy humane animal husbandry practices (4) building stable, productive humus soils. If we all come together and act now—action guided by biological-organic experts— change is possible faster than anyone thinks. While it might take a farmer seven years and more of trial-and-error attempting to convert dead soil to living, fertile, biologically active soil on his or her own, farms can achieve conversion to regenerative organic-biodynamic farming in as few as two to three years with the assistance of an experienced biodynamic/organic consultant.

Conscious Agriculture: An Introduction

Conscious agriculture understands that our systems of food production must align with Nature. Consciousness in all endeavors reflects our highest human potential to evolve—to reach complete or total awareness of the individual. Life's deepest wisdom can be learned through participating in agriculture, even gardening, which cooperates with nature and life. The science of soil life, the knowledge of the soil food web, the correct use of biology, and the collective wisdom of humankind are required to achieve conscious agriculture. We find this potential in biology-driven regenerative agriculture or biodynamic agriculture.

In stark contrast with chemically based agriculture's obsession with producing maximum crop yields. Producing clean, healthy, and nutritious food is achievable on a global scale with the will to accomplish this. It takes farming that is in harmony with nature, utilizes good farm management methods, and blends traditional wisdom with modern scientific observation. It also supports the economic well-being of the farmer: the independent farmer will thrive when no longer at the economic mercy of profits for the narrow interests of agricultural marketing firms and global corporations (food conglomerates) with their unrestrained waste of non-renewable resources. The United Nations has declared that the future of food production lies with small-scale organic and sustainable farming. A two-year study proved that these methods are also capable of—and must—feeding the world's populations. Not GMO crops and large-scale, chemical-based industrial agriculture presently supplying the food system.

The corporate naysayers will argue and fight to maintain the status quo. But if they win out, it is us—the citizens of the world—who will suffer and pay the price for inaction. Those who advocate for nature, the environment, and our planet must speak out loudly and consistently. Environmental youth activist Greta Thunberg from Sweden has thrown down this gauntlet:

Build back better. Blah, blah, blah. Green economy. Blah blah blah. Net-zero by 2050. Blah, blah, blah. This is all we hear from our so-called leaders. Words that sound great but so far have not led to action. Our hopes and ambitions drown in their empty promises...They've now had 30 years of blah, blah, blah, and where has that led us? We can still turn this around —it is entirely possible. It will take immediate, drastic annual emission reductions. But not if things go on like today. Our leaders' intentional lack of action is a betrayal toward all present and future generations. We can no longer let the people in power decide what is politically possible. We can no longer let the people in power decide what hope is. Hope is not passive. Hope is not blah, blah, blah. Hope is telling the truth. Hope is taking action. And hope always comes from the people.

If you find Greta's words are even a little exaggerated, it is time for you to wake up. Nothing is changing. In the 1970s, coal burning represented 38% of our energy consumption; in 2022, coal is still 37% of global energy production and energy needs have increased substantially. It has been all blah, blah, blah.

Can we eat our way out of a global crisis in agriculture? Can we eat our way out of climate change? Proponents of several well-managed organic agricultural systems emphatically answer yes! It will involve several key conscious agriculture methods to drive this real-world change—complex methods that already exist in science today: utilizing cover crops; heavy mulching; lowering soil pH for weed management; sequestering nitrogen and carbon for the soil food web and plants; via biology, providing production fertility requirements and moisture retention-eliminating input fertilizers and pesticides; implementing planned grazing on a large scale; utilizing Keyline contour subsoiling; deep ripping; permaculture principals in farming, and more. It comes down to this fact: regenerative organic–biodynamic holistic agriculture can deliver the results desperately needed. If we care enough to embrace and implement all (or at least many) of them, alone, each one of these systems is very effective. Collectively, they harness the full power of nature to regenerate. In less than one hundred years, agricultural soils have gone from having mainly stable soil fertility to great soil destruction, an enormous, calamitous consequence of modern agricultural methods. Now, with the revival of traditional agricultural wisdom, with new and evolving worldviews, the implementation of "biology or life-driven agriculture" has arrived—if we are brave enough to demand widespread implementation for change.

The Uphill Battle Ahead of Us—A Head-On Approach

Conscious agriculture—aka biodynamic agriculture, aka regenerative-organic agriculture—is the future, powered by the knowledge of the soil food web. If we are to have a bright future, conscious agriculture will be the *only* agriculture of the future. In 2015 Whole Foods

Market declared the "organic agriculture and food of the future" was biodynamic agriculture. Largely due to rising, conscious consumer awareness, the US organic food industry grew from $43 billion in 2015 to $63.8 billion in 2023 organic food sales in the United States reached a record high for the sector. The global organic food market reached $204.62 billion in 2023 at a compound annual growth rate (CAGR) of 9.7% The global organic food industry will grow to $628 billion by 2034 at a projected 12% rate of growth. And yet, even with exponential demonstrated growth, the organic food industry is still dwarfed by the conventional, with the US falling particularly behind. Six percent of all US food sales are organic, but only 1.2% of US agricultural land is certified organic. We rely in the US on the importation of large volumes of organic products. In contrast, the EU has 8% organic in 2022 intending to reach 25-30% organic production by 2030 and Australia is growing as much as 20% organic and biodynamic food as of 2024.

We must now turn to the experts to guide us. And no, I do not mean the "experts" in big science. No so-called specialists from the agriculture industry or academic scholars can be real experts in farming, as they are not intimately familiar with growing food. Most of these educated folks do not work with the land directly or nature. On the contrary, they dominate, eliminate, and destroy nature. At times, at increasingly alarming rates, they now even attempt to bypass nature and soil altogether! It should be much more universally alarming that as we are approaching a great global agricultural crisis, very few people in charge seem to know our true peril. Our political and industrial leaders, the few influential, gifted intellectuals, and the 1% economically privileged class seem completely out of touch. We are coming to the end of our chemistry-based production as we know it as change soon approaches.

So, who is the expert in farming? The answer is right in front of us. The farmer is the real expert, of course! The individual who has hands in the soil, who grows plants, who manages livestock, who observes daily the seasonal rhythms. The successful independent farmer who has implemented a fertility program that demonstrates financial success, and who is not dependent on purchasing most of the farm's fertility inputs. That person is an expert in agriculture.

Can farmers grow the world's food without fertile healthy soils? The grass pasture-raised crops (meats), grains and cereal crops, orchard fruits, and row crops we depend upon? Absolutely not! Good, healthy, living, biologically active soils are essential, or we won't eat. And in the face of all of this, the elite, corporate-materialistic scientific few are advocating for even *more* destruction. They are essentially calling for a takeover of food production to create an even *more* lifeless, *less* vital, soilless food production system of the future. The conscious consumer cannot allow this to happen.

In one way, science is on our right side. The primary elements and minerals (the macro-nutrients) of chemical agriculture, "NPK"— N for nitrogen, P for phosphorus, K for potassium—is a doomed system. NPK is the chemical formula that modern society has been dependent upon for the last hundred years to feed our global supply of crops and plants. Current predictions by the leading agricultural professors at Oxford University, England, one of the world's top universities, is that the end of modern chemical agriculture is approaching. In this case, I believe them. We are rapidly depleting the world's supply of "P" phosphorus, and it's now just a matter of time until it is completely wiped out. We are expected to reach peak phosphorus by 2030, and in perhaps as little as one hundred years, phosphorus supplies will be exhausted. Are we to celebrate this change or be gripped by fear of the end of the foundation of chemistry-based agriculture approaches? Even for the most conscious consumers, this huge, impending change can be rightfully scary if we do not have faith that a superior substitute will be adequately adopted in time. I have faith, however, in our capacity for change as a species. I have faith that the course of humanity's destiny that has been so changed by the woefully misguided chemical approach to food production can—and *must*—change again. It was not necessary in the past to apply superphosphate or rock phosphorus to agricultural fields, and it will not be possible in the future when this mineral nutrient is no longer available. Zero phosphate farming will be the future of food production because it must be. We have no other choice.

Dr. Steiner offered methods—unconventional yet promising for rejuvenating soil. Proven methods of feeding soil and not crops without the repeated and continual annual application of phosphorus or other outside mineral elements. Farms in Sweden, Germany, America, Italy, the United Kingdom, and Australia have documented track records of sustainable and economically competitive food production for decades that didn't rely on conventional, chemical agricultural inputs. (I.e., using very small amounts of plant substances in homeopathic doses to activate soil biology, to correct and balance scarce earth elements.) We can fittingly refer to these as non-materialistic or spiritual methods of farming. In line with the wisdom, we inherited so long ago, consciously creating a farm organism and integrating livestock are the only methods and solutions that have ever truly worked. Partnering or aligning with nature is the only real and sustainable future for agriculture. Why are all farmers not yet fully onboard? Unfortunately, too many feel it would be professional suicide to give credibility and dedication to a method of agriculture not yet blessed or sanctioned by the dogma of materialistic science and their corporate profiting overlords. The day of reckoning will arrive for bad science based on false theories and misled natural science assumptions. Through the wider holistic study of agriculture, as Dr. Steiner encouraged, it is possible to discover the deepest mysteries of life. To discover, learn, and truly know the full potential of growing, living things, the totality of nature and life must be consciously considered.

Studying and practicing biodynamic agriculture affords invaluable insights into the living world of agriculture. We will explore the topics introduced throughout this first chapter in great depth in the chapters that come, to help you arrive at those insights yourself. Do you currently still believe that applying megatons of mineral substances is necessary to grow production crops to scale? Examining the results of biodynamic farming successes globally will expose those assumptions for the materialistic, narrow-minded mistakes they are. Yes, it is well known that phosphorus is an essential, precious nutrient to plant growth and production—but it does not need to be used in this gross artificial way of the system of industrial agriculture. As always—turn back to nature. Phosphorous is found everywhere that we find flowering plants, seeds, and protein. It is found in livestock feeds, the manures of domestic livestock, and dairy and meat products. All minerals are derived from the earth's rock and phosphorus is no different. In healthy soil and a balanced ecosystem, soil microorganisms and plant roots will mine and liberate needed minerals from rock. The planet has been doing this for millennia— overwhelmingly longer than we have spent disrupting its natural processes.

As Dr. Steiner indicated, phosphorus, like all minerals, needs to be recycled in a closed system established on a healthy farm organism. Good science and a practical approach to agriculture have been developed out of the biodynamic method. A closed nutrient system should be the goal of all organic and conventional agriculture. Embracing this idea will allow agricultural science to rethink and reevaluate the present materialistic first law of agriculture. The biodynamic method is a system that works with nature to restore the balance that has been thrown off.

We must be absolutely, crucially clear: the stakes *for survival* could not be higher. The widespread climate extremes that have become increasingly common in the last few decades—and seem to be rapidly escalating—have both been largely caused by gross agricultural malpractice *and* resulted in increased challenges *for* agriculture and the stable production of food. That is food security. We find ourselves in a global, agricultural catch-22. Nature, the environment, and our current civilization are not in balance at all. We either learn again how to work with nature when practicing agriculture, or we will fail to provide the food and nutrition required for human populations to sustain themselves. Well-tested and proven conscious, regenerative, biodynamic agricultural methods paved the way for returning to aligning with Nature, the only future agriculture—and our species—has. Specifically, as we will explore in detail, the

Australian Demeter Biodynamic Method (ADBM) has shown consistently exceptional results in soil regeneration and food production and must be widely understood.

The future of food *will* be the story of good agricultural practices that are in balance with nature. We have no other choice if we hope to survive. The future of food will not be the story of greed as our downfall, as protecting corporate profits over nature and natural viability. The destructive systems of the international farm-food conglomerates, bioengineering, GMOs, CAFOs, chemical feeding of plants, fake cultured meats, and toxic spraying of plants cannot be our legacy. Not when we already have the solutions ready and waiting to be put into action around the world.

2

CHALLENGES IN MODERN AGRICULTURE

WE TAKE for granted that our modern supermarkets are filled to the brim with thousands of foods. Yet few of our foods are produced locally, but we have endless exotic choices from all over the world. At what cost and footprint do these foods come to us? What cost to the environment and our planet's health? More and more people in emerging societies want the same extravagant food supermarkets that we find common in developed countries, but can we provide with a balanced eco-friendly agriculture the endless choices and volumes of food that are in demand? It seems unlikely.

The single greatest challenge facing modern agriculture is growing enough quality food to feed the world's ever-increasing populations. This involves reducing—or eliminating—hunger without further degrading the environment and the natural world that we humans depend upon. In other words, the mutually critical functions of feeding Earth's people and protecting Earth's environments are inextricably linked. Large-scale climate impact will likely continue to dictate much of how the system of agriculture must adapt to support today's world. And, to compound these problems are rising food costs for the public. Recent continued price increases are complex but can be reduced to corporate greed and manipulation of food costs. This is due to a lack of competition and control by just a few regional, national, and multi-national food conglomerates. Anti-trust laws exist in the United States to prevent this imbalance from happening, but laws are not being enforced, lobbyists see to this, and the consumer keeps seeing their food dollars shrink. International and regional food monopolies have food production, purchasing, production-processing, and distribution in their hands. Including retail sales where supermarkets are controlled by very few players, so we see trends in price gouging due to lack of competition even in the retail distribution of food.

We begin by grounding ourselves in benchmark data from trust-worthy global sources. According to the United Nations (UN), climate change refers to long-term shifts in temperatures and weather patterns, mainly caused by human activities, especially the burning of fossil fuels. The UN's Intergovernmental Panel on Climate Change (IPCC) has determined that our food system is responsible for 21-37% of all global greenhouse gas (GHG) emissions—the biggest contributing factor being the clearing of and burning of forests for the sake of increasing industrial agricultural capacity and the melting permafrost (peat bogs are now releasing huge amounts of carbon and methane) caused by the climate warming. Increasing and continual environmental degradation thus leads to food insecurity globally. It is estimated that the atmosphere presently contains 750 billion tons of carbon. The soil is estimated to hold twice as much carbon, about 1,500 billion tons of carbon. Soil cultivation and deforestation contributed about two-thirds of the atmospheric carbon or 500 billion tons. The oceans as part of the biosphere, hold most of the earth's carbon, measured in gigatons of carbon (GtC)/ The majority of carbon is stored in the oceans, much more than terrestrial plants and soil, making up the terrestrial biosphere.

Industrial agriculture's impacts are engendering a dangerous catch-22 cycle of destruction. Permafrost in the earth's northern lati-tudes is melting much too rapidly due to rapidly rising temperatures (an average of four degrees and as much as 12 degrees in far northern and southern latitudes), speeding up soil respiration. This has released enormous amounts of carbon and stored methane gas CH^4 —a potent GHG, even more threatening to atmospheric warming in the short term than carbon dioxide CO^2. This, along with the excessive burning of fossil fuels, has caused farmers to seek solutions to what they believe are new challenges utilizing rapidly drying soils, which puts greater stress on plants and causes cyclical bouts of extreme heat (drought), prolonged dryness, and flooding—posing great risks to both the planet and food security/agricultural success. The UN's Food and Agricultural Organization (FAO) puts it this way:

"High-input, resource-intensive farming systems, which have caused massive deforestation, water scarcities, soil depletion, and high levels of greenhouse gas emissions, cannot deliver sustainable food and agricultural production."

Once understood, this catastrophic pattern makes desperately obvious the dire need for more thoughtful approaches to agricultural practice. Challenges to our food security include challenges for the farmer, corporate overreach, a broken nutrient system, healing the soil, a doomed reliance on artificial chemistry, pollution and poisoning of agricultural soils, and more. The way we produce and consume food will either alleviate or exacerbate these conditions. We must reject partial solutions, ineffective solutions, and pretend solutions. We must accept the very real consequences of industrial agriculture to life on Earth. It is time to wake up.

Challenge A: Challenges for the Farmer

Farmers in the United States and globally today are waging a devastating uphill battle.

According to the National Farmers Union in the US, the farmer's share of the food dollar is, on average, below 20% of retail sales. That means more than 80% of our food dollars go to the middlemen—

food brokers, handlers, processors, and retailers. How will farmers grow your food if economically driven out of business? Many farmers have seen little or no increase in their wholesale produce prices for decades for both conventionally grown and organically grown foods. Yet, food costs keep increasing. Who is profiting when food costs increase? The same middlemen, the big food conglomerates. In India, it is estimated that conventional farmers may receive as little as 1-5% of the retail food dollars they produce. In contrast, traditional organic peasant farmers earn ten times these amounts, selling directly in the local food economy.

This state of affairs begs a question that demands an answer: Who controls food production in America today and in other nations, if not the food producers? We are facing a situation of corporate land ownership. Today, less than 1% of agricultural lands are now owned by small family farmers. Billionaires—especially Bill Gates—and celebrities now own more agricultural land than farmer-growers do, making society's elite the agricultural landlords of the modern age. Does this not harken back to the fief system of the medieval period? Is a return to that system of land ownership already here? Decades of this trend of corporate land ownership have singlehandedly destroyed traditional farming culture and farm communities. These social issues represent the corporate takeover of our food system. (Not to mention the fact that we have paved over millions of acres of the country's most valuable fertile lands with development at an industrial scale with more industrial parks, parking lots, golf courses, shopping malls, and cemeteries all show our culture's lack of vision for a desirable future!) Add to these misuses of prime farmland the willful, abusive, and misguided application of municipal sludge and industrial sludge (now discovered to have been filled with toxic forever chemicals PFAS). It is estimated that as much as 20,000,000 prime farmland acres are now contaminated, and farm water sources, due to the ill-thought-through practice of applying industrial and municipal sludge to farmlands. At farmers' expense especially, we have been poor stewards of the Earth's arable soils.

The awareness of farm worker exploitation must become more

common knowledge, in the EU, China, and North America where affluent societies depend upon workers from underdeveloped societies. Workers are essential for manual labor for specialty crops, fruits, nuts, vegetables, nurseries, and flowers. The corporate overlords retail food chains, middlemen, or wholesalers these buyers are only concerned with the bottom line and cheaper food. This greed mentality squarely falls on the backs of exploited immigrant farm workers and forced prison labor in China. Few farmers, associations, organizations, and conscious food handlers are concerned with fair trade issues and farm worker's rights, fair pay, and legal status in most countries where their services are depended upon. The abuse and unfair practices are again the cost of mass-produced cheap food and the modern industrialized food industry.

Challenge B: Corporate Overreach

Building on this trend of corporate land ownership, another headlining challenge facing our agricultural system today is over-control by this corporate tier more broadly. There are a handful of transnational corporations—the ten largest being Nestle, Archer Daniels Midland, Cargill, Sysco Corp., JBS, George Weston, Tyson Foods and Danone, PepsiCo, and Mondelez—that control our global food supply. These corporations have the power to manipulate every aspect of the food system. From food production and distribution to food workers' labor conditions, to influence in lobbying governments and legislatures, and more. With so much power and money in the hands of so few global food conglomerates, even more disaster is looming. In addition, four multinational corporations, Dow-Dupont, Syngenta, Bayer-Monsanto, and Limagrain control two-thirds of the global seed market, of course, three of these four are also pesticide-chemical producers with 3,000,000 tons of the deadly chemicals used on crops annually. These are hybrid seeds, showing in many cases half or less of the nutrients of heirloom seed varieties, due to selection for considerations of storage, color, and size but not for nutrition.

This harmful oversight does not only apply to the financial aspect of modern agricultural systems. It also applies to the way science is wielded across the food production ecosystem. While chemistry-based, petrochemical farming is not sustainable, this system is perpetuated and promoted by farming and agricultural academics, "experts" wittingly or unwittingly are essentially serving as the salesforce of the petrochemical agri-business giants that dominate world food production.

Ultimately, it comes down to the fact that these are human beings wielding this undue, harmful influence. Humans are the only real problem and only humans can solve these problems. In the face of the alternative agricultural movement and its many proven regenerative systems, the puppets of the agribusiness industry—the "bought and paid for" academics—still always attempt to argue that organic agriculture cannot feed growing human populations, or even that organic food is not justified. These arguments are cherry-picked or incomplete at best, outright false, or intentionally misleading at worst.

Challenge C: A Broken Nutrient System

Another great challenge facing modern agriculture today is the broken nutrient cycle. Plants require, at minimum, 18 essential nutrients for healthy growth. Humans and animals require a minimum of 33 to 99 essential nutrients to achieve and maintain health. And yet, in artificial agribusiness, we feed plants only three elements (NPK) and then chase plant deficiencies and disease with applications of other minerals. How is this upside-down equation supposed to support health and life's functions? It is common sense that this kind of chemistry system is not tenable.

Excessive and wasteful irrigation is another facet of challenged plant nutrition under an industrial model of plant growth. Crop flooding and the over-irrigation, used on many crops results in washing down plant nutrients into the subsoils and increasing salinity in soils (salinization), which nothing will now grow. In California's San Joaquin Valley and Australia's Goulburn Valley alone, we are at risk of fully, permanently losing two million acres of prized agricultural land due to such artificially imposed salinization.

Challenge D: Healing the Soil

Considered more broadly, a view into agriculture's broken nutrient system becomes a more overarching phenomenon of broken soil. After centuries of harmoniously working the natural land, it may seem difficult to believe that catastrophic, widespread soil degradation has occurred in only the last one hundred years—but the industrial-chemical agricultural system has managed to achieve this great destruction.

The excessive use of water-soluble nitrogen fertilizers is destroying agricultural soils. Each year, more and more synthetic nitrogen fertilizers are used, *as of 2023 it is estimated that 109.7 million metric tons* of nitrogen soil fertilizers [1] are now used to grow humanity's food. And, in 2023-2024 estimates are 111.6 million metric tons will be used to grow food grown via modern chemical agriculture. That is 99% of agricultural land globally is poisoned with nitrogen fertilizer. This artificial process extracts the carbon humus inherent in good soils. The application of ammonia nitrites leaves these soils hardened, compacted, and more and more dead. As soon as these processes begin, it is only a matter of time before soils collapse completely, having been leeched of nutrients, becoming entirely structureless, dead soils with little or no active soil biology. And the devastation doesn't stop there. Topsoil losses (due to erosion from excessive large-scale farming) and carbon-humus loss (due to mining the soil without replenishing humus and plant residue-carbon/potassium) cause more death to soil biology. For decades, this pollution has been destroying coastal sea life and local fishing economies. Nitrogen and phosphorus runoff from destroyed soils has created a great dead zone in the Gulf of Mexico larger even than many countries. It would not be unfair to consider this destruction as a war against nature and the planet. All wars are destructive to human life, farm livestock, and the land, as seen in the Russian War upon Ukraine. This ongoing war against our planet's soil fertility is a commensurate example of the destruction of nature on a massive scale not to mention losses to human life. Is there a death wish from those in positions of power? Will society never learn? How can we return to live in peace, to treat each other and the earth with the dignity we deserve?

The U.S. Department of Agriculture (USDA) has made a devastating prediction. They have announced that current data trends indicate we will see the loss of all topsoil—that's all-humus (carbon) soils —in America by 2050-60. The United Nations has similarly announced that by 2060, 18% of the earth, its arable cultivated land surface used for agriculture will lose all topsoil. For decades already, on average, ten tons of topsoil have been lost per acre per year across North America, and we're not alone. Across Ukraine, Poland, France, Italy, and Germany—countries that have historically been agricultural *bread baskets* for decades, if not centuries; lands with deep, fertile, clay and chernozem black humus soils developed in temperate forests over millennia—conventional chemical agriculture and its decades of heavy cultivation and nitrogen fertilizer use has begun to burn up and exhaust the reserve stores of carbon-based humus soils that were once believed to be inexhaustible.

Other aspects and practices of industrial agriculture are damaging to soils as well. Modern large-scale farm implements—the steel plow and rototillers—are destructive to the soil life and soil structure. (A farmer can now kill the soil, or soil life, and destroy soil structure with a plow!) Soil can also be killed via over-cultivation and with the application of chemical sprays.

To begin the important process of healing Earth's soils, soil biology must be restored along with deep subsoiling (or ripping) and keyline design has been a very effective method in Australia. Deep subsoiling/ripping is very beneficial in beginning the processes of proper soil re-structuring, soil redevelopment, and rainwater retention. The importance of learning from and widely adopting remediating, healing soil regeneration practices like these cannot be over-emphasized. As Mahatma Gandhi said, "To forget to dig the earth and tend the soil is to forget ourselves."

Challenge E: A Doomed Reliance on Artificial Chemistry

If all of these challenges—challenges to the farmer, challenges of big corporate interference, challenges to Earth's natural system of nutrients and its soils in particular—the very industrialization process behind these factors is dooming itself along with our planet. In another one hundred years, the supply of the mineral phosphorus will be exhausted. Phosphorous is a valuable, non-renewable material, and today, in the interest of industrial agricultural practice, it is being mined on a gargantuan scale. Once it is inevitably gone, the recent historical systems of agriculture will, like the dinosaurs, become extinct.

Phosphorus occurs naturally in rock and soil. It is a critical nutrient for plant growth. For centuries, old-world farmers added phosphorus to their fields as recycled farm manure to maintain and boost harvests. According to the US Geological Survey (USGS), world consumption of phosphorus pentoxide fertilizers (super-phosphate) increased to 48.2 million tons (Megatons/Mt) in 2019, up from 43.7 million Mt in 2015. The world's phosphate rock production increased gradually as well, rising from 223 million Mt in 2015 to 255 million Mt in 2019. Most of this increase comes from the expansion of existing mines in Morocco, Africa, and a relatively new mine in Saudi Arabia. The cost of mining these megatons of phosphorus has doubled in the last ten years, continuing to raise food costs around the globe by an estimated 10%. These trends and impacts are, of course, under current, unchanged practices, expected to continue.

This widescale global depletion is resulting in chaos across the farming world. Farmers have been advised to purchase tons of phosphorus compounds over thirty years. In some cases, in a best-case scenario, these farmers will see 25-45% of this purchased phosphorous utilized by their crops. However, due to the existing extent of soil damage and ruin, many may see as little as 1-3% utilization of these expensive resources taken up by their crops, with the remaining 97% or more of the purchased phosphorus remaining in the farmers' subsoil. What a waste of money, for the farmer, and of natural resources, for our world—and the wastes ripple outwards and outwards when you consider how the widescale mining and artificial application of phosphorous plays out within the full agricultural ecosystem. Imagine how these statistics scale up when applied to a global practice. Imagine the cost to the environment from the diesel-driven super shiploads and trainloads and millions of truckloads of rock phosphorus mined and hauled to farms globally. Most conventional factory farms and industrial organic farms in every state in America—and every nation of the world—import phosphorus from Morocco, a country that supplies 85% of the world's agricultural phosphorus. More and more hundreds of billions of U.S. dollars, Euros, Pounds, and Asian currencies will be sent to Morocco and Saudi Arabia for this non-renewable mineral that chemical-based agriculture is dependent upon.

All these bad habits are simply accepted as the normal cost of input agriculture doing business as usual. But soon, we will have to change the way this works—we won't have a choice. The inability to grow food with chemistry inputs is foreseen and predicted with the end of the non-renewable mineral phosphorus in sight. An Oxford University professor of agriculture has predicted this will result in "the end of modern agriculture as we know it." As others say, the next great agricultural crisis is looming!

Partial Solutions, Ineffective Solutions, and Pretend Solutions

We know solutions to these challenges exist. But some of the solutions being attempted right now are partial solutions at best—and pretend solutions at worst.

Each time the soil is turned by aggressive industrial cultivation practices, half of the soil organisms are destroyed. While they might be able to recover naturally over time, they cannot and will not recover in the face of repeated, continuous application of chemicals and repeated aggressive cultivation. These excessive cultivation practices are driving many younger farmers and gardeners to advocate for "no-till" farming or "earthworm cultivation" farming and gardening. This is reactionary, of course, to the abuse of soil cultivation. And this is only a partial solution. No-till agriculture is not a complete and correct solution to the important problems in large-scale industrial agriculture because they are unsustainable. Via other methods, proper cultivation or conscious disturbances to the soil are essential to good soil management.

Unfortunately, the industrial scheme suggests the use of "state-of-the-art technologies" as solutions to current challenges—including precision farming, remote sensing (e.g., through drones or satellites) to detect disease or nutrient deficiency, and robotics (e.g., for weed control)—which actually only compound these challenges. No matter how you spin it, any reliance on farming technology is very much the opposite of ecological, regenerative, organic-biodynamic farming. Pitched as scientific advancements, GMOs and even the hybridization of seed or livestock are, scientific errors! Monstrosities even. Evidence of backstepping science and reversals of wisdom and nature's intelligence. With "solutions" such as these, nature's intelligence is never addressed; only "man's" intelligence and ability to manipulate nature are ascribed value. These false "solutions" are simply the next step in the same cycle of materialistic thinking that has caused the environmental and health crises we presently have. A conscious, experienced, well-educated, and observant farmer who is sympathetic to nature, who is aligned with nature, and developed inner perception is the real solution to our most pressing challenges of food production today—not the use of more and greater technology.

In a gross abuse of power, some now posit *alternatives* to natural food as a potential agricultural path forward. Scientists are now culturing and growing lab-grown meat tissue intended for the mass fast-food market. They even claim these kinds of unnatural monstrosities as being healthier for the planet, for livestock, and for human nutrition. This scheme is a sci-fi nightmare in its worst-case scenario.

Meat substitutes or substitute meat from plant-based peas or soy are not new. However, the techniques involved in their creation are becoming increasingly grotesque. One California producer, Impossible Foods, has added *heme*, a cultured component of blood, to their plant-based meat. In Israel, the company Future Meat is lab-culturing manufactured meat and growing meat and fat cells in a hi-tech facility, requiring high energy costs and producing waste byproducts not disclosed. Even if we set aside the inhumanity of these practices, on the level of logistical consequences, the sanitation challenges caused by the large-scale production of cell culture technology—growing meat or flesh—are highly expensive to address. Food contamination and food poisoning are inevitable on a large scale. These sci-fi solutions are absolutely not the solutions to our present-day challenges. We simply need a system that allows us to go back to producing better quality and eco-friendly food, for both plants and animals.

(As a side note, human populations in Asia, Africa, South America, and France eat insects, because insects are efficient at producing protein from plant matter. Growing insects for human consumption is another proposed solution to producing protein food for humans and reducing proteins from livestock. Do beetle burgers sound appetizing to you? Do grub snack bars? Mealworm breakfast foods? These are not the ideal vision of the future of food production but may become more and more of a reality with a true lack of vision for the future of food.)

Industrial agriculture proposes many multifaceted, catch-22 solution strategies that involve the use of more insecticides, herbicides, and fungicides. It is being preached that these expensive toxic inputs are necessary to improve yields and relative yield stability. But the fact is that for one hundred years the USDA has kept records of crop yields and losses, and this data reveals that regardless of our growing system, nature has always taken on average 20%, before and after the use of insecticides, herbicides, or fungicides. The public has been deceived that chemistry increases production. It is a fact that all crops can be grown ecologically, organically, and biodynamically. An eco-strategy or agroecology is the preferred approach to not just our most pressing global challenges, but our food production processes and relationship with nature overall. Our solution-oriented focus should be on natural strategies that get us back to a place of good farm stewardship, enhanced plant nutrition, environmental protection, seed sovereignty, and soil health. Back to the roots of agriculture, and our cultural inheritance. The path forward is agriculture aligning with Nature.

Faux-Sustainability

There is a lot of talk in the food world today about *sustainability*. The definition of *sustainable* is "being capable of being sustained"—not being sustained in part. The food and chemical conglomerates have prostituted the definition of sustainability, to the extent that the word retains little meaning in the world of agriculture today. The term is being conveniently applied and sold by corporate agri-business to strategies that only promote the use of more Agrochemicals. This is a public propaganda program to justify toxic, polluting, non-sustainable chemistry-based agriculture—our industry's failed system.

In the world of agribusiness, "sustainable" is a marketing term. Consider these sad truths: the word "sustainable" has been hijacked —trademarked—by the corporate agriculture conglomerate Monsanto Corporation. Results from a quick internet search for "sustainable agriculture" reveal the top five sites are corporate chemical companies and agricultural universities (falsely) touting the wonderful and idealist tenants of a sustainability agenda. Today, the concept of sustainable agriculture is nothing more than a gobbledygook of well-sounding goals, a meaningless soup of words that don't match the reality and actions of "sustainable agriculture's" toxic deception.

Even more than the term *organic agriculture*, the meaning of sustainable agriculture has been diluted and "sold" to the unsuspecting grain farming community and the public as a good thing, as a positive alternative to chemical agriculture. When in reality so-dubbed sustainability products of the agribusiness world are simply wolves dressed in sheep's clothing. Don't feel bad—most of America's well-intending citizens have been duped right along with you. The public, wholegrain bakers, and farming communities are caught in the web of corporate deception around the marketing of the sustainability term. Many promoters and consumers alike can't, for the most part, articulate what sustainable agriculture is! That is a paradox.

The truth is always in the details. Just follow the money. As a marketing program, the favorable connotation of the sustainability label sells corporate products and increases profits. The details or definitions are worthless when Monsanto promotes "sustainable farming" to unsuspecting grain farmers in the northwest by helping these farmers become "no-till" farmers, and their true corporate sales strategy all the while is to promote greater and greater use of the deadly herbicide Round-Up to kill weeds that no-till does not address. (These days, sales and use of Round-Up are up. In 2017 the California State legislature officially classified Round-Up as a carcinogen or cancer-causing herbicide.) America again shows the benefits to the few and sells the short end of the stick to the many: the natural environment and the everyday consumer. This is a win-lose reality.

These alarming facts and deceptions are a challenge to all conscious-thinking individuals. Hopefully, an informed and courageous society will awaken to the brazen corporate manipulation and misrepresentation of the term "sustainable." My suggestion to the individual consumer? No action is too small. Buy and use certified organic grain, cereals, and bakery products, thus bypassing the corporate deception of "sustainably grown" altogether. We will go into great depth on the topic of organic agriculture in the next chapter.

The Real Cost to Life

Costs to nature, the environment, and human health are never factored into the industrial agricultural system decision-making process. And yet, the costs are very real and already here. In America, we are seeing skyrocketing healthcare costs in a correlative connection with skyrocketing environmental costs. The poor nutrition and poor health of much of the American public reflect the underlying cost and bottom line of materialistic thinking and chemical-driven agriculture. We reap or end up paying—upfront or eventually—for all that we sow.

Will the sobering statistics convince you? The negative effects that mass-produced, chemical-based food, GMO foods, highly processed foodstuffs, and junk food have had on America's health and well-being are tragic. The American Medical Association (AMA) has predicted that young people twenty-five and under will have a one-in-two cancer rate in their lifetimes, up from the Baby Boomer generation's one-in-three cancer rate. Is this acceptable? In what world is a 50% cancer rate among the generation of today's youth—our children and our grandchildren—acceptable?

The Food and Agriculture Organization of the United Nations (FAO) approves pesticides as part of its projects and most of these chemicals (supplied by Syngenta, the Chinese chemical giant) are being used in southern countries. Many of the most toxic chemicals banned in the US and the EU are sold and used in Brazil and Africa for example. "Paraquat". Farmer poising is far too common, a study published in 2020 found about 1.6 million cases of human pesticide poisoning in Europe, 8 million in South America, and over 115 million in Africa in a single year. These chemicals are often funded by the United Nations FAO and government programs supposedly supporting small farmers. This is a part of the tragic cost of growing food with chemical agriculture. What are the real costs to families and societies from this toxic agriculture? In Africa one in four poisonings are fatal, resulting in over 28 million deaths to farmers annually on one continent. Now consider the public poisoning from eating these toxic sprayed foods year in and year out. It is estimated that 10,000 different chemicals and over 6 million metric tons of pesticides are used on, food crops annually. Here are some of the real costs of modern food production systems. Government subsidies in the US and the United Nations support GMO and chemical agriculture, at farmers and the public's pearl. We can add catastrophic medical bills, and environmental degradation that are and will cost the general public the consumer.

A Call to Face These Challenges Head On

A radical rethinking is required in modern agriculture today. It is no longer acceptable to deplete and exhaust non-renewable resources. This message has been repeated, again and again, with few listening, and only a few governments implementing known eco-agricultural solutions.

Our greatest, overarching challenge is our inability to reject our stubborn, collective, ongoing destructive habits toward the environment—those that continue, in real-time, catastrophic toxic pollution of our food, our soils, and water systems. Allow me to be necessarily blunt, and very clear: *We have no acceptable future if we do not embrace regenerative organic-biodynamic practices. If we do not—and soon— humanity will be collectively lost. If we further destroy nature's abundance and vitality, we destroy ourselves.*

One way or another, it all comes down to agriculture. If the environmental costs and degradation of nature don't kill us outright—via extreme climate change events, for example—they will starve us out as a human species by making it impossible for us to produce enough quality food to sustain life. Will the collective public get their heads out of the sand? Will the consumer who wields the power of their purchasing dollar (currency) become sufficiently educated in the urgent need to understand how food quality, food production, and its distribution, soil health, and farm health are interconnected with our environmental health, species restoration, and planet-wide ecological restoration are all connected?

Defending Real Solutions

So, what is the solution in the face of these challenges and high stakes? Carbon farming, conscious regenerative, ecological organic, and biodynamic farming. That is the only solution capable of fixing the great damage we have wrought.

It is widely recognized that, compared to conventional, industrial agriculture, organic agriculture generally has a positive effect on a range of environmental factors: biodiversity, useful carbon sequestration, soil quality and hydrology, erosion mitigation, and the reduction of climate extremes. So, what is the resistance these solutions face?

Once again, it comes down to money. Specifically, the assertion that organic farming is more expensive for the farmer to produce our crops and organic products are more expensive for the consumer to purchase. And this is simply not the case! The full picture tells an importantly different story—one in which conscious, organic agriculture is not only the only viable environmental approach but also a sounder economic one in the long run.

Today, the costs of all industrial agriculture inputs are going up: those for synthetic fertilizers, powdered rock minerals, diesel fuel, and poisonous chemicals. When this happens, food prices rise. In other words, industrial agriculture is not the sounder economic system it is often touted to be. Who profits when food costs rise? It is not the farmer or grower of our food, but the corporations pulling the industrial puppet strings. And it is the consumer who always pays for rising artificial costs every time we buy conventionally grown foods. The wool is being pulled over our eyes. Because the true costs of continuing industrialization are never calculated in the real cost of food, chemical agriculture continues to be (falsely) touted as the smarter financial choice, *and* the additional costs to the environment and human health are conveniently ignored. Factored into the bottom-line conventional chemical agriculture is far too expensive for human societies, the environment, and our climate overall.

The complainers—those paid-for advocates of industrial agriculture—tell us that organic farms produce 20-40% less volume, on average, than conventional production, and that this means farmlands growing organic food are less productive models we can't afford. What an absurd assertion! These flawed arguments conveniently forget to mention that the net profit or bottom line for the farmer growing with organic and biodynamic practices is, in most cases greater than conventional farmers. Because eco-farmers and regenerative farmers do not purchase all the mineral and chemical inputs required of industrial agriculture, their net profits ultimately end up being greater than those of conventional farmers. Often, biological farmers produce less volume or less water-filled produce, but their bottom line is better because they are producing higher-quality food.

The real, dangerous problem is not this flawed thinking alone. It is that these misinformed or misdirecting "experts" continue to offer biased, untrue information to businesses and the public at large, as a purposeful attempt to create controversy around organic agriculture. The same deceitful tactics they use to falsely tear organic farming down are the same they use to falsely prop chemical farming up. For example, "scientific advancements" is one term used to hide the sales strategy and manipulation of companies bent on selling more inputs to farmers. These unnecessary inputs fatten corporate profits and further squeeze the already narrow profit margins of conventional farmers and growers if they have any net profits at all. Throughout the current state of this industry, we find corporate deception at every turn.

We must ask ourselves: what do we as a species truly need right now? Do we need more volume and more commodity crops to manufacture lab food? Or do we need better nutrition per acre and nutrient-dense real food? The battle cry of corporate chemistry-driven agriculture is to endlessly increase crop yields and produce more volume! Produce more calories! Of course, it is much more sensible to grow less volume of more nutritious nutrient-dense food. Conscious farming supports livestock health and farm fertility, stopping the pollution, exploitation, poisoning of farm workers, and rape of the environment. And, through several key methods, it does so in a way that is more efficient than traditional methods today.

Consider two linked, prime examples: the organic practices of crop rotations and cover cropping. Organic farmers rotate crops and fields. Cover crops and crop rotations improve soil fertility and are destructive to pests and weeds—a real economic gain for organic farmers because higher fertility inputs create greater growth rates of crops. To be clear, these are not new methods. Crop rotation is a basic practice and is even required in organic farming certification programs. But even this is promoted by industrial farming advocates as a negative feature of organic farming practice! This tactic of criticizing the fundamentals of organic production is just another false flag from the chemical-based agriculturists who claim cover crops and rotating crops lessen productivity. What is the industrial agricultural alternative to these basic, sensible practices? Applying more water-soluble, nitrogen-producing, unhealthy, often toxic-laden chemicals to crops in bulk—practices that are completely wasteful and environmentally polluting, often causing crops and farmers to become dependent on pesticides. The public health concerns are real and yet sidelined by the cheering for more production of more nutritionally void commodity crops. You tell me: which sounds like the more economically viable approach?

In short, biological organic farming is the only real choice available to us if we care about our future. The only argument against it—that it costs more money in the long run—is a false one. So why are we, as a global community, so stuck? Why do we stick our heads in the sand? The good news is this: important people are starting to wake up and drive real change forward.

Food Fraud

Another challenge facing modern civilization is food fraud, estimated annually to be a 40-billion-dollar deception. Most of the food fraud violations have been found coming out of China. And an emerging solution is surprisingly coming from China. Food fraud is the intentional deception of customers about the quality or content of food products for financial gain. It can include: substituting ingredients, adding undeclared ingredients or materials, concealing poor quality ingredients, mislabeling, counterfeiting, and document forgery.

Today blockchain systems are being developed to trace food from farm to table. Said, to increase transparency and accountability. This would not be necessary if the food was not a global system fraught by big money interests. Local food and its production, the certification of organic, and Demeter biodynamic foods of course already address the food fraud issue. Every step of certified food is traceable, audited, and verified by inspectors. But, in all things, if there is an opportunity for some unscrupulous people to distort, manipulate, or completely defraud the public this system of blockchain would not be necessary. Tracking your food origins and its trail to your hands seems like an emerging trend. It is not without costs in fact, it will become another multi-billion-dollar industry and yes, the consumer will pay for it!

In England (UK), for some decades all food has been tracked nationwide (The Carbon Footprint of food or GHG emissions) from the origin country to the supermarket. The carbon footprint tracking system is becoming more sophisticated with a national goal to reduce by half 50% GHG emissions on food by 2030. This system gives the consumer information on the real carbon footprint, an environ-mental consideration when purchasing any food. It is absurd that in the US the food industry claims it is impractical and not possible. Here again, big corporations can control our government policies in the US. It is of course, possible and could be implemented if the public demanded this traceability. Presently, 25-35% of GHG emis-sions globally are generated by our food production.

The World is Starting to Wake Up

It is shocking, that many vegans and vegetarians do not see the forest for the trees. Most people are meat eaters. Most people have blood types O and B that benefit from animal proteins. Although all people could be vegetarian, it is very unlikely as much as the vegan idealist dream of this. Many people's health would be weakened without animal products in their diets. Cultural customs, habits, and family traditions cannot be changed overnight if at all. Yeah, for a more plant-based diet for much of humanity. But is this a reality? The issues are much bigger and more complex than just advocating for a plant-based diet. If organic regenerative and biological agriculture are not the top priority for vegans and vegetarians then they are still eating, toxic laden, poor nutritional foods and still supporting the industrial agriculture system and not supporting the environment nor halting climate extremes. Consider the 20% of global grassland used for grazing livestock. It is balanced and practical to feed humanity with grass-fed livestock and improve natural ecosystems. First, we must stop utilizing grains to feed livestock to produce meat and dairy production in CAFO operations. We must feed humans with all the grain now being fed to livestock this action would make a major shift in human food consumption.

Dr. Cynthia Rosenzweig is the senior research scientist at NASA's Goddard Institute for Space Studies (GISS), where she heads the Climate Impact Group, studying the relationship between climate change and food. She pioneered NASA's study of climate change impacts on agriculture and cities. She co-founded the Agricultural Model Intercomparison and Improvement Project (AgMIP), a major international collaborative effort to assess the state of global agricultural modeling, to understand climate impacts on the agricultural sector, and enhance adaptation capacity, as it pertains to food security, in developing and developed countries.

Since 1988, the Intergovernmental Panel on Climate Change (IPCC) has studied food security. The organization now recommends that we waste less food and consume fewer animal products or none. In an ideal world, they now say, we would predominantly eat more plant-based diets.

These are examples of how the spotlight is starting to shift, to shine on and illuminate the real truths of our current, precarious situation as a global species.

The threat to life as we know it on this planet is at stake. Take the enormous, manmade dead zone in the Gulf of Mexico as a terrifying, representative example of the impact and cost we are reaping via our wasteful, industrial, life-killing agriculture. This madness is entirely avoidable. Many life forms essential to a healthy environment and our perpetual food supply are being threatened ever greater each year. When will these destructive practices stop? Will the life that is sacred to agriculture and ecology be recognized as sacred to life itself?

The greatest loss brought to bear upon humanity and livestock is the loss of nutrition and the nutritional values in our foods, as the soil food web is reduced and destroyed by destructive agricultural practices and poisonous Agri chemistry. The "web of life" essential to supporting the healthy growth and development of plants and animals—humans included—has been disrupted. This may at first sight seem like a minor issue, but it is of the greatest significance! When a seed (cereal grains for example) is of low fertility, or a plant is unhealthy or imbalanced, life cannot reproduce itself. A sick or unhealthy plant growing from soil that has been disrupted, destroyed, or even made lifeless cannot reproduce generation after generation—can no longer carry forward the foundation of life-giving natural forces.

It can be difficult for those tightly bound to materialistic thinking to grasp the profound point made here. It's not just about nutrients and their quantities or qualities. (Although their presence or lack thereof does tell us something about quality.) It comes down to the basic nutritional qualities in the seed and in the plant's nutritional foundation which is key to the finer understanding of a plant's nutritional value. Essentially, because of the damage we have wrought, we must now always ask ourselves: are the plant's vital life forces—its trace minerals, enzymes, etheric forces, prana, or chi—active and alive in the seed?

When soil life is disrupted or destroyed, a plant planted in that soil is unable to reach its full potential to develop—because key natural relationships that form a complete and healthy plant are missing. Missing nutrients (missing minerals) result in a plant missing its complete nutritional profile, it's immunity to disease, lacking vitality, and this has a relation to the entire cosmos. Yes—our planet, our plants, and even us humans are connected and must relate to a cosmos or wider universe. These are not simple ideas to grasp at first, but when seen from a true depth they reveal themselves as profound knowledge.

In addition to seed vitality, seed ownership has become a major issue. Who controls or owns the seeds to plant life on our planet today? Seed control numbers among today's numerous pressing challenges in agriculture—which are now a part of the greater challenges of climate extremes, environmental pollution, species loss, food security, food quality, farmland ownership, and a pathetic farmer's share of the food dollar. We now know that life involves relationships, and how we relate to nature, to humanity—and to each other—will certainly determine our future. A plant's seeds involve how it relates to the whole of its environment or surroundings. This is "the nutrition of connectedness"—not a quantity to be measured, but a value measure of a plant's quality of life. Literally, can an unhealthy plant reproduce life generation after generation? No, it will fail to do so because it no longer has its connection to life itself. This same principle applies equally to livestock and human fertility and health as well.

It all comes down to soil. Life's diversity is nowhere more evident than in the earth's soils! As much as half of all species on earth are believed to be living in the soil, and most of the complexity and diversity of soil's life forms are still unknown. Soils are individualities and are unique. Soil, plants, cultures, and ecosystem health are linked in ways we are still attempting to understand. The unique mix of soil microbes can determine which plants can grow and thrive in a particular environment, and which can't survive! We have much to learn as a species, as intelligent, spiritual, sentient beings. We risk so much by devastating our global soil stores in the catastrophic way we are currently embracing.

It is terrifying that we have arrived at this crossroads of life versus death. While it may seem that humankind has achieved the upper hand over nature, we risk so much by crossing dangerous, unnatural lines. Our inaction and the collective unconsciousness in today's societies—in the face of such a crossroads—is a human lack of will to change, evolve, or grow. But the time is now. If we don't act presently, our inaction may be our real downfall. Because whether or not we act, we will soon be faced with many emerging worldwide challenges around food, agriculture, and the environment. We must embrace, conscious agriculture, our only hope is to begin to right our collective wrongs as quickly as possible. We better hope we have begun to actively respond when nature comes inevitably calling.

3

WHY ORGANIC AGRICULTURE IS COMPROMISED

Unfortunately, in its current state and iteration, the organic agriculture movement has been coopted and compromised.

The admirable ideals that the first biological-organic growers envisioned in the 1950s and 1960s have been diluted by practical–industrial-driven objectives to expand the organic industry. But today, the organic farming industry has become a $63 billion industry in the U.S. [1], global organic food market size was estimated at $183.35 billion in 2022 and is expected to be worth around $546.97 billion by 2032 [2], —and with that economic scale has come the tainting influence of corporate abuse. As we know, the interests of large conglomerate food corporations and government bureaucracies are often not the interests of consumers and the environment, no matter what agricultural approach they are supporting. Essentially, as soon as big academia, corporations, and government got involved, the future of pure food, true "organic" was doomed.

It all comes down to the system of organic agriculture being nationally and internationally managed, like a business. Corporate advisors of the National Organic Standards Board (NOSB) consider and make recommendations on a wide range of issues involving the production, handling, and processing of organic products—including certifying "exceptions" to organic standards. Who are the board members of the NOSB? Well since 90% of all larger organic brands have been bought up by a handful of large food conglomerates, who appoint the board members to advise the USDA Organic Certification programs overseers. Today, NOSB decisions are often more interested in industry growth and increased sales marketing choices for profits over the goals of purity and quality that gave rise to the initial organic movement in the first place. Because of both this dangerous influence *and* increasingly bad science, the certified organic farming and food industry is following in the footsteps of the materialistic, chemistry-based agriculture industry to which it was supposed to be a remedial solution. The public has been short-changed from the ideal of organic agriculture.

Are the ideals and goals of organic agriculture off-base? Absolutely not! When pared back to its essentials, organic certification is very important as the first step to more conscious agriculture and a regenerative organic-biodynamic future. We must simply work to understand what has gone wrong and corrupted today's organic agriculture movement and seek out a better path forward— to finally arrive at a truly authentic organic method, to biological agriculture growing in good fertile soil, to the nature-based process-agricultural methods of the future.

Today's Organic Lie

The breadcrumb trail back to the root of the big organic deception is not easy for the distracted and manipulated, on-the-go consumer to follow. Money, power, and influence of the corporate food conglomerates have corrupted the original National Organic Program. The first organic farmers who began the organic movement now see nothing but greenwash for the USDA's "faux-organic" program.

Consider this: today, the term *organic* is not even considered a health claim. This is the backward mentality of bureaucrats. Today, for the most part, organic agriculture is mass-produced organic, or "industrial organic." This is the real standard for certified organic food.

The industrial, chemistry-based NPK approach to growing food exists in opposition to the touted standards and goals of a biological-organic approach to organic farming. But make no mistake. The way that organic agriculture works today, it is just as much of a primarily chemistry-driven system as its industrial-conventional counterpart. Importing non-synthetic "approved organic ingredients" OMRI rock minerals, plus (i.e. non-organic manures, feathers, and blood meal) from industrial factory farms to provide fertility for certified organic farms? This is a joke. The perceived need for and embracing water-soluble fertility is a huge compromise on the part of USDA's Certified Organic system.

In truth, the environmental footprint being perpetrated by today's system of "organic" agriculture comes at great environmental cost. Beyond the non-synthetic NPK fertilizers being used today on organic farms, energy from fossil fuels is required to transport those NPK "organic" OMRI-approved fertilizers to the industrial-organic farms. Supposed standards of authentic, ethical farming are not compatible with a factory process of assembling parts. Organic agriculture has become just another industrial system where the people in charge believe that the larger the scale, the more quantity gets produced, the better. (On the contrary, real food quality is produced in small to medium-scale farming that can be managed in balance with nature and respect diversity and balance between plants and animals.) It is a grave error that so much of conventional organic agriculture today is driven by the very unsustainable, unrealistic, materialistic, chemistry-based systems and goals it claims to oppose.

Given all this deception, it requires a truly informed consumer to be able to see through the marketing jargon and recognize the real profound differences between materialistic, science-based food production and the spirit-life-bio-based food system. A biological life science approach to agriculture includes, by definition, more life and mankind as part of the co-creation approach to food production.

Striving for the True Organic Promise

Organic agriculture promises food purity and health. For the most part, it seems that *organic* can only deliver marginally healthier food than conventional, pesticide-laden, toxic foods. But we must also contend with biased information gathering at the source.

Whenever the question of the merits of conventional versus certified organic food is raised, a false argument or equivalency is always promoted in the media. It purports that organic food, as it is produced today, is nutritionally superior. Studies comparing nutritional values between today's conventional and organic food offerings and products have shown marginal or no nutritional differences between the two categories. This would suggest that the public's perception of organic food as more nutritious than conventionally grown foods is false—that the public has been intentionally misled.

Yes—the public has first and repeatedly demanded "clean food" food with no synthetic nitrogen, insecticides, herbicides, or fungicides in their food. What are we getting from today's organically produced food? However, at the same time, few if any of the studies reporting to compare certified organic vs. conventionally grown foods were truly objective. Privately sponsored nutritional lab analyses from several biodynamic and organic farms have shown significantly higher quantities of vitamins and minerals in their food product. However, none of these nutritional analysis panels have been published for scientific review due to a complete and real bias that prohibits any scientist or institution from publishing such results. So, it has been shown that when truly biologically grown organic food is objectively studied, it is verified and proven that it does offer nutritional superiority over chemically grown organic food.

The real organic mandate must be that all crops must be grown in biologically active organic fertile soil. Plants must be in full contact with the earth and nourished by the natural biological activities of that soil for uncompromised nutritional value. Research into the marvelously complex soil micro-biome reveals the vital ecological processes that support natural, non-chemical food production. These processes provide the necessary fertility for soil and crops without importing non-organic inputs to organic farms.

True or real organic food—that which is biologically grown in healthy, fertile soil—will always be superior in quality and nutrition to chemistry-grown conventional and chemistry-based organic foods. This is due to greater soil humus, available cation soil minerals, an emphasis on soil and plant health, highly developed plant feeder root systems in composted, pastured, and rotated fields, on-farm nutrient recycling practices, and the use of pastured livestock. Deep-rooting green manures, cover crops, more perennial plants, and grazed pastures must be included within broadly based crop rotations to enhance soil fertility and biological diversity.

A self-contained fertile soil should be maintained principally with farm-derived materials and made into compost. This avoids bringing in unwanted or polluted material from industrial sources, invasive species, and pathogens. Thus, fertility is created and maintained within the farm.

Biological activity is developed by growing green manures, cover crops, perennials, and grazed pastures. Healthy plants feed soil biology, the microbiome of soil. Included within the strategy of broadly based crop rotations, soil fertility and biodiversity, are enhanced. The greater the variety of plants and animals on the farm, the more stable the system. Of course, these methods are all biodynamic principles. Prevention of imbalances is the key to a farm's health and productivity. Biologically active fertile soil and healthy plants induce resistance to pests and diseases in the crops. Including pastured livestock is the pinnacle of good agricultural development.

The Role and Importance of Accurate Information

So, how can you tell the real biological organic from industrial organic? Biodynamic Demeter certification standards deliver real insights into the integrity, health, quality nutrition, and purity of food produced from farms that make organic claims, as well as a greater public awareness of variations in biodynamic quality. Consumers in

Australia, Germany, Sweden, and many other countries in Europe—where these standards are already widely in place—recognize the higher quality of biodynamic foods over most organic food choices. But in America, where the marketing of biodynamic products has not been widely implemented, this recognition does not yet exist for most retailers, organic consumers in the greater public. Very powerful marketing forces in America have managed to dilute, distract, and confuse the public with information that doesn't attest to a food product's true organic and biodynamic status. Here, the Cornucopia Institute—a genuinely admirable non-profit watchdog organization within the organic industry—does the best work informing the public and everyday consumers of true organic quality products—but there is a great deal of work ahead of us.

A few decades ago, a big, hyped-up deal was made about implementing and enforcing food labeling standards in America. The results are now no more than a joke, as the labels we now have on our foods provide no real nutritional information of significant value. Do these food labels give data on essential nutrients, vitamins, minerals, or essential oils to the public? Absolutely not! At best, some include information on iron or calcium, but most only list numbers for calories, sugar, and sodium. Most B vitamins added to refined flour products are synthetic (petrochemical derived) and have no real nutritional value, but the consumer is led to falsely believe they are receiving essential nutrients in their flour and bakery products. What information would a genuinely useful food label provide—given what the average consumer needs to know about chemicals being artificially added to the products they consume? To adequately raise the right kind of awareness, labels should emphasize magnesium and potassium levels in foods—these nutrients offset the high levels of phosphorus in all Americans' diets today. Levels of essential nutrients and macro and micronutrients should be on labels as well. In addition, if raising awareness is the true goal, meaningful public education on nutrition should be provided in schools and via public service announcements, so a truly informed public can adequately understand and contextualize the label information that is provided today.

Certifying Organic: Going Beyond Food

Food that is truly organically produced requires organic standards to be met and maintained across every step of the food production ecosystem.

Consider this. Very few organic consumers would feel good about or give their approval for the use of industrial feedlot cattle manure, or commercial-industrial dairy and chicken manure, being used to fertilize their certified organic vegetables. These feedlot animals are being fed GMO feeds (corn and soy products) as significant portions of their diet. Overarching organic oversight organizations greenlight these practices—these CAFO feedlot manures are the source of OMRI (Organic Materials Review Institute) "organic approved" manure—but how is this organic? How can any rational decision maker qualify the manure and feathers from two-hundred and fifty thousand chickens living in a single factory farm building—a confinement industrial operation: a CAFO-consuming, GMO feed, egg-laying factory—as organic? How does this make factory-farmed chicken manure OMRI "organic approved" used on USDA Certified Organic farms? This is another big compromise in organic production.

It's a contradiction in terms that organic materials or manures can both come from conventional factory farms and be used to produce certified organic produce and crops. Although these manures are from "non-synthetic origins" (considered, by scientific definition, as *organic from carbon-based sources*) they are anything, but "certified organic", yet those in charge of the organic certification programs give them the go-ahead now for more than three decades.

Proponents of this hypocritical arrangement and system argue that this compromise is necessary to supply enough nitrogen-phosphorus fertility (animal manure) believed to be required for organic food production. But is this non-organic manure necessary to grow organic food? No, it is not! That is a materialistic, chemistry-based misconception perpetrated by those who do not understand or study soil biology and biological farming. This system is only necessary because of an industrial chemical approach to growing organic food. A true biological approach to growing regenerative organic and biodynamic foods does not need such non-organic inputs. Biological-organic and biodynamic regenerative farms develop soil biology—and raise their own livestock thus, producing fertility—right on the farm, reducing and eliminating the importation of non-organic manures. Soilless hydroponic systems can not be considered certified organic but since 2017 they are.

It requires elevated thinking and rationale to become a successful biological-organic farmer. Growing organic certified foods biologically—work which, in most cases, prohibits the use of manures from non-organically certified sources—is a more complex challenge. But it's worth it. It's necessary. The compromises currently being allowed, by a far-from-perfect OMRI system, are too high of a price to pay. (And all this misuse comes right from the very system instituted to allow materials and manufacturers to provide materials for the USDA Certified Organic scheme!)

There are so many examples of ways in which the USDA certification scheme is broken and compromised. For example: we may need to reconsider organic certification standards when a vegan or vegetarian is happily purchasing and eating their organic greens and vegetables grown using blood meal from non-organic industrial feedlot livestock slaughterhouses. Yes—this blood meal is OMRI-approved for your certified organic spinach, salad, and vegetable dishes. Yum! This can be hard to swallow for food purists.

Another case: approved at least three times since 1990, the National Organic Standards Board (NOSB) has approved the use of prohibited "synthetic" amino acid methionine in all poultry and livestock feeds. *Certified organic standards prohibit any use of synthetic ingredients or materials.* But still, this not-so-public deception has been promoted and approved by the NOSB as an exception now for thirty years! The given justification is without this exception, the organic poultry industry would pay for expensive livestock feeds that would raise organic poultry prices by at least 25% or more. Would you pay for organic eggs at $12 per dozen or organic chicken at $10 per pound? These are the hard choices being made. Sadly, in the face of economic headwinds like this, these kinds of exceptions are becoming the rule.

There is some more discouraging news. In 2017, the U.S. Department of Agriculture took a step toward increasing the production of organic grains for livestock consumption—which has not kept pace with demand—by launching a program to certify farmland that growers are *in the process* of converting to organic operation. According to the Organic Trade Association, an industry group, obtaining certification under the program allows grain farmers to sell "organic" products raised in accordance with organic guidelines, after only one year instead of the three-year minimum requirement for organic certification, for higher prices than conventionally grown goods. That helps growers cover the loss of production that is often associated with transitioning to organic farming—an important preliminary step. But, unfortunately, another compromise to organic quality.

Where Organic is Headed: Continued Steps Back and New Steps Forward

A genuine organic solution and system is in real demand. Even in the face of all the headwinds intended to distract and mislead them, individual consumers are waking up. Demand for organic foods is strong, with no sign of slowing, as consumers are increasingly seeking organic products considered healthier choices than conventional foods. According to the Organic Trade Association (OTA), organic food sales in the United States in 2023 hit a record $63.8 billion, with total organic sales reaching $69.7 billion. The OTA's 2024 Organic Industry Survey also found that organic produce accounted for more than 15% of all fruit and vegetable sales in the U.S. While there are considerable problems with the food being qualified as organic today, this is still a hopeful trend. Certified Organic foods are essential to the future of clean and healthier food.

The USDA has proposed that the organic system "facilitate the investment in transitional agriculture through a consistent set of rules, and ultimately support the continued growth of organic agriculture." There is no sugarcoating it—this is no simple task. Farmers desiring to become certified organic must grow crops for three years without using prohibited substances, such as genetically modified seeds, artificial nitrogen, Round-up, and synthetic pesticides. Essentially, for those three years in which those farmers convert their farmland to organic production, they must follow the same regulations as those who have already been fully certified without yet realizing the economic benefits of the certification.

If and when an amended certification system is implemented, grain farmers (growing wheat) stand to benefit the most. This is because they are the sector in which demand has most outpaced supply. For at least three decades, there has been a severe supply bottleneck as the high demand for certified organic grains has grown and grown. In never-ending cycles, farmers are often grappling with very low prices for conventional grains because of a global supply glut in the market. Because organic grains have seen heavy demand —driving up their price—it has become very expensive to feed organically raised livestock organic grains if the grain is available at all. Similarly, there has been a high demand for organic bread and cereals for humans. Here is the unfortunate compromise being made to address this need: grain farmers must only prove they have been following the guidelines for organic production for one year to be certified as transitioning their land, not the three-year standard established for all other organic agriculture. Agents accredited by the USDA will verify compliance. The new program, producing "organic" grain for livestock, does not provide standards for labeling food grown on farms that are in the process of transitioning to organic. This deceptive scheme is deemed necessary to get more farmers into organic production. And so once again, sadly, quality is compromised for the organic consumer.

Large-scale organic produce production is incentivized on a volume-weight basis—that is, organic produce is paid for by size, weight, and cosmetic appearance. Unfortunately, true measures of quality—nutritional value, flavor, and environmental impact—are not the most important criteria. This is exactly, the same system utilized for conventionally grown produce. Isn't organic supposed to be about quality food?

Importantly, when true organic standards are not upheld, we get large or watery produce from the water-soluble nitrogen chemical approach to industrial food growing. Not only does this not support healthy soil or healthy plants. It also means that the weight we pay for in produce is water weight—more carbon! More water in our food does not equate to more nutrients, flavor, or quality. If organic growers use water-soluble nitrogen (also found in blood meal) and non-composted animal manures to "feed plants," your health and the environment are compromised. (To be certain organic certification requires the use of composted animal manures, but the hot and aerobic compost process required to be used, in certified organic agriculture, is designed to "kill pathogens" and it is not about creating stabilized nitrogen compounds (humus), which is achieved with warm or cold composting that requires more time to develop the best quality compost full of beneficial biology). Compost produced by high-heat 140-180 degrees F, is not quality compost.

Quality is not about larger size, water-laden produce. Quality is achieved when our food is nutrient-dense with minerals, vitamins, antioxidants, phytonutrients, essential oils, trace elements, and amino acids. That is quality in nutrition! Let us not forget, that organic agriculture is presumed to promise more nutritious food. The IFOAM definition of organic farming is a touchstone by which we in the industry must operate and live:

> *Organic Agriculture is a production system that sustains the health of soils, ecosystems, and people. It relies on ecological processes, biodiversity, and cycles adapted to local conditions, rather than the use of inputs with adverse effects. Organic Agriculture combines tradition, innovation, and science to benefit the shared environment and promote fair relationships and good quality of life for all involved.*

Today, a small percentage of biodynamic regenerative-organic farmers are paving the way. With their small-scale, biologically driven organic farms, they are truly producing and delivering higher quality—for the exclusive promise and goal of biodynamic food production is food grown for higher nutritional quality. Biodynamic agriculture sets the highest standard for clean and nutritious food products, produced on balanced and healthy farms, driven primarily by biology (rather than by chemistry alone). Importantly, the biodynamic approach to organic agriculture does not accept the same compromises the traditional organic movement absorbs. The biodynamic movement understands it is necessary, and essential that we have a growing and thriving "organic industry" that prohibits the use of GMOs, synthetic Petrochemicals, and many more inhumane practices in the management and raising of livestock. Growing food in healthy soil is fundamental to quality organic and all-biodynamic food.

The conscious, informed consumer has a role to play in driving this mode of agriculture forward. We can—and must! —demand better organic. Biological-organic, regenerative-organic, and biodynamic must be the choices of the informed, conscious consumer now and in the future. The watchdog Cornucopia Institute has been the best advocate for integrity in organic production and compliance. The informed consumer will not only wish to avoid or reduce poisons in their food; they must also be motivated by the environmental, climate impact of agriculture and nutritional values on a broader, global scale.

By now, it should be clear that the same reductionist science of the chemical approach to plant cultivation and agriculture largely predominates the prevailing "science" of faux-organic agriculture today. These kinds of false schemes should seem familiar to most American consumers. This kind of compromised model is eerily similar to why American health insurance is always going to be a failure: because it is co-dependent on a disease approach to medicine instead of emphasizing prevention. Building health and prevention of disease is achievable. Building humus soils and organic matter is the foundation of a farm's health and is achievable.

The nature of organic cultivation and the key to plant health and nutrition essentially boils down to feeding the soil (natural) instead of feeding the plant directly (artificial). The overarching biodynamic or regenerative organic goal has always been to create balance and health on the farm, which leads to both healing the planet and preventing future disease. A treatment-based health approach—i.e., chasing disease—is no longer necessary when health and balance are achieved on a farm, because, at the end of the day, health very much comes down to what we put in our bodies. A balanced and healthy farm organism will produce healthy plants, nutritious healthy food, healthy livestock, and, in turn, healthy consumers. This is the objective and legacy of regenerative-organic and biodynamic agriculture. Developing truly self-sustaining farms is the real goal.

In its current, predominant form, today's system of organic agriculture is a middle-ground system at best. Yes, it meets the lowest bars of natural standards, but the movement's willingness to accept (at best) or embrace (at worst) the importation of genetically modified by-products, (GMOs), with polluted soils, and chemical residues from conventional agricultural systems is untenable. Other groups and movements have sprung up to attempt to address the disillusionment with USDA's certification of organic food. The Real Organic Project is one such organization that insists on traditional organic methods of growing food. Growing food in the soil, not hydroponically. Raising livestock on pasture, not in confinement. Real Organic remains exactly what organic was always intended to be in the face of these widespread organic compromises. Yet, the high ground is held by the Biodynamic Demeter certification and the new Regenerative-Organic Alliance (ROA) certification (to be implemented soon). These systems strive to achieve farms' fertility needs from within the farm property, reducing or, eventually, eliminating the importation of 80-100% of a farm's fertility.

Today, it is still true that the cost of organic certification has become perceived as too costly for thousands of smaller, local farmers who would choose certification if this process was simplified and made more affordable. Put another way, the barriers imposed by important, necessary costs and processes inherent to the true organic system have summarily reduced the number of organic farms and acreage operating in North America. Unfortunately, we must accept that having the organic certification process be truly free of compromise—for both our human health and the health and continued livability of the planet—results in a capacity for the organic operation being geared toward larger farms and corporate entities over local, small, family-owned farmers. Still, public safety and the burden of food safety and quality control should not fall exclusively on organic agriculture. Conventional farms and their poisonous system should be tracked, monitored, and reported to the authorities and the public. So many uphill battles left to be waged! A national map should be online for the public and farmers to see the lands where PFAS and industrial and municipal sludge have been applied. These farmlands should be tested, and lab reports available to the public at government expense, not at the farmer's expense.

It is an optimistic sign that today, 58.8 million hectares—over 145 million acres—are now in certified organic production globally, but this represents only 1% of cultivated agricultural land. We can meet the challenges in modern agriculture—but to do so, we must improve, upgrade, and maintain our agreed-upon and enforced organic standards. Ultimately, we need hundreds of millions more global agricultural acres to be developed as conscious, biological-organic farms, and biodynamic farms. If we achieve that, we can restore farm fertility, close the nutrient cycle, and grow more nutrient-dense foods. We can reverse global topsoil losses. We can sequester excessive atmospheric carbon! But none of these achievable outcomes will be achieved, if we do not embrace this improved path forward toward biological farming, and soon.

4
____

WHAT IS BIODYNAMICS? REGENERATIVE ORGANIC AGRICULTURE

BIODYNAMICS IS A TRANSFORMATIVE, regenerative, and self-sustaining practice. A holistic approach to agriculture in all its facets—one that aligns with nature and strikes true harmony with nature. This is what modern consumers are seeking; what modern culture is asking for; what the planet today needs to be able to continue sustaining life. And biodynamics is ready to go! The precepts of the approach have been identified and proven and are already being practiced. What are we waiting for?

> *Indeed Rudolph Steiner was one of the first people in the modern era to recognize explicitly the principles of inter-connectedness in relation to farming and to describe the links between the fertility of the soil and the health of plants, animals, and people.*
> Prince Charles of Wales (Presently, King Charles III of England)

While you may not have heard of it before this book, don't be fooled. Biodynamics is actually the oldest method of conscious and organic agriculture, with its introductory lectures presented in 1924. *The Agriculture Course*, as it is known, was given by Dr. Rudolph Steiner—a scientist; a lecturer; as some academics refer to him, a mystic. Steiner's agricultural lectures were requested by individuals in his anthroposophical circle. That is, his community who was interested in his leading perspective on the intersections of agriculture, science, and health. The eight-lecture course was the first introduction to an ecological and alternative agricultural idea ever presented. Steiner described this course as "a course of lectures containing what there is to be said regarding agriculture from an Anthroposophical point of view." He said:

My subject was the nature of the products supplied by agriculture and the conditions under which these products grow. The aim of these lectures was to arrive at such practical ideas concerning agriculture as should combine with what has already been gained through practical insight and modern scientific experiments with spiritually scientific considerations of the subject. (Steiner 1924)

Biodynamics refers, quite simply, to bio (life) and dynamics (the activities, forces, and/or processes of life and the cosmos). It is a holistic system—the complete approach to an eco-agriculture practice.

Think of it this way: in biodynamics, a farm or property is referred to as a "farm organism." A living organism. A closed, self-contained ecosystem. A biodynamic farm will strive to produce everything it needs on its own farm property, from seed to animal feeds to produce. Animal manures are used in biodynamic compost (using special biodynamic preparations). Promoting the propagation of diverse plant species is a goal of biodynamics. Because, ultimately, working and in cooperation with nature, the goal of the biodynamic farmer is to create and rejuvenate humus soils. This allows for the regeneration of habitats and diverse species (e.g. soil with maximum earthworm populations!), the improved health of plants and livestock, and the economic productivity of the farm.

In many ways, the biodynamic farmer and gardener can be seen as nature's assistant—aligning with and for nature while producing healthy, amazing farm products. The use of the biodynamic method supports healing and health, because a biodynamic farm is a farm in balance, producing the most nutritious foods possible.

Biodynamics is the promise of regenerative organic agriculture. Many people believed that organic agriculture would be the answer to healthy food and farms, but the organic movement has been compromised. All along, it was always the uncompromised Demeter certification standard that offered true hope for the future. The biodynamic method is the only path forward capable of delivering the promise of healthy agriculture, our shared future on Earth.

Biodynamic Preparations

All biodynamic agriculture is set apart from other organic farming systems by its use of, fittingly, biodynamic preparations. The biodynamic preparations are specially prepared materials used in small quantities, activated, and applied to soil and plants. These preparations are the critical answer to regenerating soil in a very short time and on a large scale. Small amounts of special materials can rejuvenate life in soils. Critically, they involve forces, not chemistry or material input substances. How is this accomplished?

Dynamic change takes place via forces. Quantum consciousness and new thinking involve forces. The biodynamic preparations have the power to create change because they are akin to a wave or force, much like the power of the seed, to start a new life and this can bring about sweeping change. A new idea or a powerful idea, which time has come, can cause a much-needed change. The forces of change start with the kernel of thought that can build and swell to create a momentum large enough to move mountains. And, of course, consciousness can shift and change minds equally. To see a revolution in agricultural thinking and behaviors we need to act and manifest these forces of life into being.

In poor soil mineral feeding microbes (bacteria) alone can create soil biomass when properly made and applied biodynamic preparations are used. Soil microbes both bacteria and fungi are activated and proliferate under ideal conditions, living and dying in multiple generations in a single growing season, leaving their dead bodies behind as necromass (micro aggregates). In soil, necromass is a significicant and long-lasting part of soil organic carbon, which is the main carbon pool on land. Dead microbes and plant matter are key elements in soil fertility. Necromass is a vital source of nutrients and persistent carbon in soils. It is stabilized by bonding to the soil's mineral matrix and aggregating. In one acre of biologically active organic soil, there can be a minimum of one ton of microbes and even more than two tons per acre depending on soil conditions, moisture, and food sources. The decaying bodies of soil bacteria, mycorrhizal fungi, nitrogen-fixing bacteria, and plant material contribute to the soil biomass. Mostly bacteria produce important soil aggregates (glues, microbial necromass) and nutrients (soil carbon). Plant roots and their exudates along with other soil life proceed regenerating soil structure. Soil fungi make strands that bind the micro aggregates to then make larger macro aggregates in the soil, increasing further soil structure, this opens the soil for greater aeration and better soil respiration. Improved soil fertility and greater life in the soil are the results of this activated soil-web of life.

Well-studied soil biologists may appreciate the processes at work in good soils but would be amazed at the ability of the biodynamic preparations to activate the life forces. Individuals educated based on academia's standard scientific frameworks can be challenged to comprehend or accept this agricultural spiritual science approach—except perhaps for some physicists and quantum physicists—but it is a timeworn approach that harkens back to the very origins of agricultural methods on our shared planet. Life begets life! Specifically: six herbal, humus-like substances known as preparations (502-507) are referred to as "compost preparations"; two preparations known as "horn-humus" (500) begin as horn manure; one known as "horn-silica" (501) is made from quartz crystal; one known as "herb horsetail" (508) comes from equestrian arvensia. These substances have the power to transform life in soil and with plants.

Proper activation and application of the biodynamic preparations foster the proliferation of biology via greater, improved soil life.

Horn-silica 501 enhances plant photosynthesis. Horn-silica 501 enhances plant light metabolism, invigorating growth, and health. When activated and applied properly, the biodynamic horn-humus 500 communicates with soil life and plants fostering the proliferation of life. Preparations are used in homeopathic quantities. Horsetail 508 helps control fungi and molds.

Within the category of biodynamic preparations, there exist some varying biodynamic methods. While the preparations known as 500 to 508 are utilized in all biodynamic methods being practiced, one method may be necessarily particular to a country, region, or continent. The Australian Demeter Biodynamic Method, for example, varies somewhat from the German method, and both methods are practiced in North America, Italy, France, and elsewhere.

All biodynamic preparations have specific uses. Their activation and application may need to vary in order to be most effective, depending on the practitioner, nation, soil conditions, and geographic region. Most biodynamic farmers following the Australian Demeter Biodynamic method activate or stir their preparations in copper stirring tanks or pots. In Germany and some European countries, wooden barrels are commonly used. In America and the United Kingdom, practitioners most commonly use stainless steel, ceramic crocks, wooden barrels, and copper tanks. In New Zealand and in North America, stirring systems made of concrete composites (agricultural flow forms) were developed and used to activate the preparations. Most of the biodynamic preparations are made by groups of biodynamic farmers or professional preparation makers, who are usually farmers as well.

In industrial agriculture farms, mineral imbalances are "corrected" by artificially importing mineral inputs. As we have examined, this approach creates far more problems than it could ever solve. However, on biodynamic farms, mineral imbalances are corrected via the holistic approach of the biodynamic compost preparations. In this process, fertility is created or remedied by means of developing soil biology, which can correct most soil imbalances—as improbable as this may seem. Materialistic-thinking soil scientists who have been traditionally trained in plant and soil chemistry are especially challenged to accept this as truth, but the proof is in the results. Expanding our minds beyond the limits of artificial science is one of the first steps along the journey to a healed relationship with nature and agricultural soils.

A Self-Sustaining System

The biodynamic method has been heralded, by some, as *agriculture of the heart*. In some circles, it is even being called the new organic! In truth, biodynamics is the original conscious, biology-driven agricultural system—cooperating with the intelligence of nature to produce healthy and nutritious food. One could say this is harvesting the results of Nature's intelligence.

Biodynamic farms avoid all synthetic inputs and, most importantly, use natural plant and mineral materials for creating and maintaining the farms' health and fertility. They are not indebted to cheating or compromising policies for their success. They are "organic" and so much more.

Applying the biodynamic method to farming operations leads to the development of a self-sustained farming system—one that does not need to rely on continuous external, artificial inputs to meet its natural fertility needs. Instead, the farm organisms responsible for the self-fertility on a biodynamic farm are its own farm livestock, most importantly the cow, honeybees, and earthworms. Aka, *the fertility trio*. This makes such logical, natural sense, it's almost inconceivable how this isn't a natural, strategic component of fertility for all modes of agriculture: of the top ninety most consumed foods, seventy are pollinated by honeybees. Furthermore, biodynamics and all biological growing systems are driven by soil biology, and knowledge of the soil food web informs us that the earthworm is the important animal most responsible for developing topsoil. Earthworm castings and their calcium capsules, along with decomposing organic matter, are incorporated into the earth while earthworm tunnels aerate the soil. With maximum earthworm populations, ten tons of castings can be produced in just one season on a single acre of land. The castings equal to a thin dime's worth spread over an acre of soil. And, in twenty years the entire top eight inches of soil can be digested and turned by a healthy earthworm population.

Many years ago, Benjamin Franklin stated that "an ounce of prevention is worth a pound of cure." He was referring, in that case, to *human* health, but his wise concept applies just as fittingly to agriculture as well. Biodynamic farmers and gardeners focus their attention on building healthy humus soils, thus preventing plant and animal stresses and disease. This can be understood as working with the order and organization of life or life forces.

The Metaphysics of Biodynamics

Demeter is the Greek Goddess of life and the harvest. Fittingly, Demeter has become the international name and symbol for biodynamic agriculture worldwide. Independent national Demeter certifying organizations –the premier eco-agricultural certification authority—work with biodynamic farms to certify their actions meet and exceed international organic standards and requirements.

Henning Sehmsdorf, Ph.D.—retired professor of Nordic Mythology at Washington State University—is a biodynamic farmer on Lopez Island, Washington. His family farm is S & S Homestead. He wisely states that there is a metaphysical component to biodynamic agriculture practice:

> *"Biodynamic management involves intuitive and meditative practices as much as scientific observation and practical applications. In contrast to conventionally produced and processed foods, biodynamically grown foods are rich in vital energy and not just in chemically identifiable nutrients."*

For some practitioners of biodynamics, this conscious agricultural perspective and practice might be experienced as a life meditation or worship. The consideration of the whole includes both the farm organism and the farm's wider environment. Biodynamics considers a *vital life force*, which may be comparable to what indigenous populations in the Americas refer to as the *great spirit* in all things—*mana*, in Polynesian tradition; *prana*, in India; *chi*, in China. This vital, "ethereal" life force is understood as physical and metaphysical, material and spiritual.

"Spirit is never without matter." "Matter is never without Spirit." Rudolph Steiner.

Those agricultural practitioners who ascribe to this philosophy believe in two symbiotic truths: that nature knows how to heal itself and can correct its imbalances, and that biodynamic methods assist and support nature's innate healing capabilities by working to create a balanced, independent farm organism that operates in harmony with the surrounding natural world.

Farming takes place within nature, not outside of nature in isolation from her wider influences. Farming can't and should not be taken out of the sphere of nature's influences, which are often ignored or forgotten in industrial-scale farming and high-tech, biotech, hydroponic, or vertical growing operations. The biodynamic method is involved in building the farm's health and transformation of its soils. While healing the earth is a noble idea, as conscious practitioners and consumers we must remember that we are small players in a vast natural universe and ecosystem. Biodynamic practitioners do not heal the earth; instead, we assist and serve nature to begin healing itself, to allow such needed healing to take place.

Biodynamic vs. Traditional Organic

Dr. Steiner, who presented the founding ideas behind the biodynamic method, one hundred years ago, has stated that we must "feed the soil, not the plant"—as in, we must focus first and foremost today on feeding the soil food web—the order of life—so that healthy soil will then, in turn, feed the plant, provide good nutrition, and prevent illnesses. Put another way, biodynamic farmers and growers do not focus their energy and resources on treating disease; rather, we focus on building health instead, putting our energy and attention into what we want to manifest instead of what we don't. When imbalances in the soil and the environment are corrected, health is a natural result. Building health and developing prevention is the *long-term* strategy for fostering healthy soil, preventing plant disease, and correcting nutrient imbalances. Stresses, health imbalances, and disease, in plants and livestock are generally no longer found on well-established biodynamic farms and properties.

Biodynamics also embraces the principles of holistic management, a planned management system of land reclamation. This is an eco-agricultural system, founded by Alan Savory, a wildlife biologist from Zimbabwe, Africa. Savory's holistic management system, taught by the Savory Institute, is dedicated to the ecological restoration of the world's arid lands and degraded grasslands through the strategic, planned rotational grazing of domestic livestock, ruminants, and cattle primarily.

In short, the biodynamic method supports natural earth health and fertility. It promotes a true state of self-sufficient sustainability, which leads to a productive carbon sequestration program. These goals of the biodynamic farm are the answers to challenges that, for a long time, many believed could be solved by traditional organic agriculture.

Biodynamic agriculture's balance between practical farming practice and a spiritual-scientific worldview is a large part of what makes it the *real answer* that "organic" agriculture cannot achieve. But there are additional glaring contrasts between the biodynamic and organic systems as well, including the biodynamic focus on building and regenerating soil fertility as humus soils; fostering and creating plant and animal health; growing food for nutrition; striving to reduce the volume of imported materials to the farm; promoting biodiversity; implementing carbon sequestration. Essentially, creating and achieving true, self-sufficient sustainability. These goals are the outstanding practices and goals of biodynamic practice that truly set the system apart. (It cannot be ignored that many of these goals of biodynamics were, originally, touted as goals of traditional organic agriculture systems. The qualities you thought you were getting when you purchased certified organic food are to be found with Demeter-certified biodynamically produced food!) Unfortunately, as most organic agriculture today has become a very large-scale, industrialized, and chemical-driven system, the gap between industrial-organic and biodynamic agriculture becomes ever greater with time.

Regenerative Organic Agriculture

The Regenerative Organic Alliance (ROA) is making strides to become the next organic certification program that is biologically driven and often leads farmers and practitioners to biodynamic practice. Perhaps toward Demeter certification, in addition, for many farmers. Soil Health Academy (SHA) is another important regenerative organic movement and educational non-profit organization located in the Midwest, North Dakota demonstrating and teaching the principles of biological "soil-centric" regenerative cultivation and land management. Then the Learning from Nature, Hill Top Farm – the Learning from Nature demonstration site and educational farm in far northern Australia, where Dr. Wendy Seabrook has a hands-on ecological teaching farm. The focus is on soil health, minimal soil cultivation, soil cover via pasture, cover crops, or mulching observing that biodiversity encourages farm health. Little or no disruption of plant roots is practiced, and livestock are used for soil health. Emphasis is on working with the cycles that support ecosystem processes. Thus, regenerative systems such as carbon farming, embrace most of the biodynamic principles. Yet, most regenerative systems have not discovered the power of the special homeopathic-like substances and processes known as biodynamic preparations.

The Shift is Beginning

It may surprise you to learn that England's King Charles III has owned an organic farm for decades—and, in the last ten years before becoming the King of England, has been using biodynamic practices on his dairy farm of 900 acres, in Wales. He's well-read on Steiner's approach to agriculture and can articulate biodynamic-sustainable agriculture concisely:

"In farming, as in gardening, I happen to believe that if you treat the land with love and respect (in particular, respect for the idea that it has an almost living soul, bound up in the mysterious, everlasting cycles of nature) then it will repay you in kind." —Prince Charles of Wales

How can farmers today go about making the transition to true biodynamic practice? By becoming a good biological regenerative farmer, and by applying biodynamic preparations. The promising reality is that many more farmers today may be practicing biodynamics than receive credit for it. The Demeter USA website lists a directory of most of the certified biodynamic farms in North America, but many of the best certified Regenerative Organic farms (ROA) also practice biodynamic methods and simply may not have applied yet for biodynamic certification. In addition, many more farms operate using biodynamic principles and materials, outside of certification. Consumers must ask for the best quality, natural food, grown on the healthiest soils with truly clean organic inputs, and these farms are going to often be biodynamic (Demeter) or biologically driven regenerative organic certified farms (ROA). These farms' produce can be discovered at the local farmers markets, or by doing your own local farm research and interviewing local eco-farmers and growers.

With thousands of successful biodynamic farms in over 62 countries, interest in biodynamic agriculture is no longer fringe. In fact, it is now mainstream agricultural practice in countries like Germany, Australia, France, England, and Sweden, nations that lead the world in biodynamic food production. In many Asian countries over the last decades numerous biodynamic farms have been developed by forward-thinking farmers. Furthermore, biodynamic food can now be found not only in natural health food stores and community-supported agriculture (CSA) farms but also commonly in natural food stores and supermarkets throughout many countries. (To find it, always look for the Demeter logo.) A person choosing to support biodynamic farming should also seek out their local CSA programs, local farmers' markets, and organic and natural food stores—especially Local Food Cooperatives, and Whole Foods Markets, which endorsed biodynamic agriculture in 2015 as the "organic-sustainable farming of the future."

Biodynamic vineyards are thriving and producing the best high-quality wines in the world. The United States (California and Oregon), Australia, France, Italy, Germany, Austria, Spain, South Africa, Mexico, Brazil, Argentina, and Chile are all biodynamic wine producers. Biodynamic wine is *good* wine that has been distinguished on the basis of quality. While only 1% of all wine on the market today is certified biodynamic, these biodynamic wines receive more than half of all major wine awards. In North America, biodynamic wines and produce can be found online, in good wine shops and Whole Food markets.

Biodynamics has been at the forefront of organic farming development, within the broader organic movement, including, for example, the system's participation in founding the International Federation of Organic Agriculture Movements (IFOAM). The biodynamic movement has been outspoken against GMOs and synthetic petrochemicals, excluding them from all Demeter-certified biodynamic agriculture from the beginning. In much of Europe, biodynamic farming and food is the dominant form of organic cultivation. With consumer support and participation, the demand for biodynamic food will soon become mainstream in North America and even more countries. For at least a decade Permaculture and Biodynamic collaboration has been ongoing as these systems can mutually learn and benefit from each other.

Biodynamic farming is the sustainable organic farming of the future. It is the preeminent -holistic perspective and approach capable of truly developing a healthy farm ecosystem that cycles essential nutrients. It is a truly self-sustaining agricultural system.

Why do most, if not all, academics and scientists within university systems and government agencies not understand the biodynamic concept and principles? The simple fact is that aside, perhaps, from quantum physics, modern materialistic sciences cannot grasp the core biodynamic tenets: the idea of subtle forces working within the biodynamic preparations; the principle that life creates and sustains life; that life cannot exist by chemistry alone; that frequency, waves, vibration, and the motion of light and sound permeate the cosmos, with all of matter and life vibrating or carrying information, from soil particles, all the way to cosmic bodies in our solar system and beyond; that all life and former life carry or hold an etheric body or life force. In reality, the insights of physics can help us understand more clearly the workings behind the biodynamic preparations.

To fully grasp Dr. Steiner's ideas—his views, language, and vocabulary—one must first study the theories of Anthroposophy. Our educational systems and culture do not, for the most part, address the wider, subtle realities of the cosmos or the spiritual worldviews that humanity has inherited. It took many years for me to completely penetrate Dr. Steiner's ideas myself. True education—or the kind of education that concerns Dr. Steiner—should be creative, interactive, experiential, and self-driven. As Nietzsche once said, "As a thinker, one should only think of self-education." And, as Louis L'Amour put it, "All education is self-education. A teacher is only a guide, to point out the way, and no school, no matter how excellent, can give you an education. What you receive is like the outlines in a child's coloring book. You must fill in the colors yourself."

5

PRINCIPLES OF BIODYNAMIC AGRICULTURE

AN INTRODUCTION TO THE AUSTRALIAN DEMETER BIODYNAMIC METHOD

NOW THAT WE have laid a foundation for understanding the biodynamic system of agriculture, we can begin to explore the concept more deeply.

As we've covered, the principles of biodynamic farming (first presented in 1924) were the first such principles of conscious agriculture. They include the basic principles of all good farming, regenerative organic farming, and gardening systems. Including soil creation and regeneration, the farm's individuality or farm organism, the farmer, a holistic system, cooperating with nature and its intelligence, growing food for nutrition and quality, and supporting healthy social connections.

The best way to deepen our understanding of the biodynamic system is to expand our understanding of each of these basic principles step by step.

Principle One: Soil Creation and Regeneration

The ultimate, primary goal of biodynamic agriculture is to develop humus soils. Creating humus soils, regenerating soil, and developing good earthworm populations are central to the biodynamic method.

Achieving this goal begins with good soil cultivation, with appropriate and limited/minimal soil disturbance. Then, subsequent steps and techniques in the biodynamic process include soil cover or cover crops (living mulch), growing pastures, and mulching soil. All this contributes to good soil structure, soil moisture retention, cooling of soils, feeding the microorganisms, and soil animal life (or the soil biome). The result is the ultimate end goal of biodynamics: stable, fertile, productive carbon-rich humus soils that are dark and rich in humic and fulvic acid and have good soil structure, high biological activity, increased soil water retention, and plant nutrient availability. Humus stores carbon, water, biology, and minerals needed for plant health. The pH of good humus soils will range from 6-6.5 to 7.-7.5.

Biodynamics produces humus soil by enhancing soil life. More specifically, creating humus-rich soils is accomplished by implementing the practical principles of good soil health, and the proper application of the biodynamic preparations. (The preparations include cow horn humus (500), biodynamic compost (502-507), and Barrel Compound (BC, 502-507). These preparations are unique concentrated humus materials used in homeopathic doses—we explore each in individual depth at the end of this chapter.) Regular practices of eco-agriculture—practices like keyline soil structuring, and deep ripping of soil with a subsoiler—greatly improve soil environments, while their biodynamic practices heal the soil. Such advantageous, holistically managed, biologically driven regenerative organic farming soil practices include the incorporation of productive pastures, properly managed grazing, proper timing of cultivation, the use of appropriate cultivation tools or farm implements, and good, low-till soil management. They have the power to bring about the transformation, regeneration, and creation of healthy soils and reap effective and efficient crop production.

It essentially comes down to achieving the right conditions for soil to become rich, productive topsoil. Achieving true nutrient cycling—thereby building up soil humus and fertility—requires that biodynamic farms maintain a balance between farm production and the exportation of farm products. Preventing soil erosion and soil compaction requires working with the weather and seasonal conditions. (Using the biodynamic calendar for the proper timing of planting and cultivation can assist plant health and humus development.)

Building earthworm populations is also especially important. Earthworms have an enormous positive impact on a farm's fertility. Each year, to create and maintain healthy soil, biodynamic farmers must aim to develop at least one million earthworms and a net ten tons of worm castings per farmed acre. We can greatly accelerate this process by growing and providing an abundance of organic matter (plant material, biomass) and producing animal manures, as the production and application of biodynamic compost supports thriving earthworm populations. And, when we incorporate manure from ruminant livestock—cattle, in particular—that provides the best quality organic matter for stimulating earthworm populations. (The production and use of biodynamic compost are desirable wherever adequate quantities of manure and compost are available. If compost is in short supply, then barrel compound (BC) made with the compost preparations (502-507) can be applied—to meet the required use of the compost preparations. For making BC a fermentation period of three to six months completes the process. Prepared 500 (horn-humus) is applied twice per year on all biodynamic farms. These combined materials and practices build humus soil and fertility while increasing earthworm populations.)

By envisioning and encouraging abundant earthworm populations, biodynamic practices are capable of yielding about twenty-three earthworms per square foot—or two hundred worms per square yard. When we achieve these conditions on our farm or garden, the soil becomes nutrient-rich, porous, crumbly, productive humus topsoil with great structure. Along these lines, let's go back to thinking about the biodynamic fertility trio—earthworms, cows, and honeybees. In addition to working with the earthworm for developing fertility below the soil, we can work with the cow for fertility on the earth and the honeybee for fertility in the air above the ground.

These soil-focused aspects of the biodynamic practice are of special importance to the conscious biodynamic farmer. Consider, as an example, Alex de Podolinsky—a small dairy farmer who taught biodynamic farming in Australia after he took over a failing biodynamic endeavor near Powelltown, Victoria. On this small dairy farm, especially, he produced cow manure for making large volumes of biodynamic preparation 500 horn-humus. Podolinsky, a soil health pioneer, introduced biodynamic farming far and wide in Australia from the 1960s until his passing in 2019. He helped to develop and refine the Australian Demeter-Biodynamic (ADBM) method.

Podolinsky was considered a controversial figure, by some, due to his outspoken focus on the biodynamic method. But he was an innovator in fact. Today, the ADBM method is widely applied throughout Australia, transforming tens of thousands of acres. In 1990, with the publishing of the important book "Secrets of The Soil", by Peter Tompkins and Christopher Bird reported that one and a half million acres were under biodynamic cultivation in Australia. This biodynamic method, a practice duplicated many times, has consistently demonstrated and proven the outstanding impacts the system can achieve concerning soil regeneration and humus development in as little as two to four years (a typically very short period for good soil development).

As of yet, the embracing and adoption of these essential biodynamic soil improvement practices are mixed at best. Yes, some farms in Italy, France, and America have had great soil results with the practice of applying the Prepared 500-horn-humus, but even some biodynamic practitioners in both Europe and America have yet to embrace the ADBM method. But, as they say, the proof is in the pudding—and we are seeing some indications that progress on this front is beginning to gain traction. Importantly (and unsurprisingly) conventional agriculture has not been able to match the results of the ADBM biodynamic practice. Even outsiders and skeptics of the biodynamic method of agriculture have shown themselves to be, in some cases, downright envious of the successful results of the ADBM Prepared 500 horn-humus method.

Principle Two: Farm Individuality or "The Farm Organism"

Creating a closed nutrient system is required to achieve and maintain a farm's health and productivity. This involves implementing biological diversity, nurturing plants annuals and perennials, seed saving, and practicing good animal husbandry—all of which goes back to the biodynamic farmer viewing the farm as a holistic system to be integrated into the wider landscape of nature.

Conceiving one's property as having a sense of place is essential. That is, the farm's individuality or identity, in the context of its unique, natural characteristics, geology, soils, microclimate, biology, and species diversity. For instance: How does the property lay in the landscape? Are its boundaries natural, or are they arbitrary and unnatural, sliced from the landscape? A biodynamic farm with a natural boundary and a healthy perimeter seeks to find and create harmony with the inclusion of fields, meadows, pastures, orchards, groves, woodlands, and or wetlands wherever possible and wherever they naturally occur. Dr. Steiner said the farmer would be paid dividends, in many ways, by this inclusion of natural landscape spaces in a farm, and he has been proven right. Diversity in the cultivation of crops, livestock, and a farm's natural environment all contribute to this fundamental principle of a healthy farm as a nutrient system.

Utilization of on-farm resources is essential on a biodynamic farm. Steiner suggested that a farm can never be truly healthy, or it would be a sick farm if it continually needed to import its fertility. On-farm resources include all carbon materials (used for mulching and compost making), cover crops, and animal manures, making/ applying biodynamic compost (the proper handling of animal manure). A focus on relationships and managing the environment to support living organisms is crucial to the biodynamic method. Utilizing the herbal compost preparations 502-507, biodynamic field spray and barrel compound support the best use of on-farm carbon and nitrogen resources.

It makes sense that all these biodynamic practice principles are connected and mutually promote each other's success. Seed saving and producing as much plant and animal propagation as possible helps to increase soil fertility and supports farm genetics and disease prevention. The core tenet of a closed-nutrient-system philosophy is that a farm knows itself best. It knows when and how to use its own materials, and it recognizes when those materials are returned to the property. (Annually, roughly 40-60% of what is produced on the farm should be returned to the farm—recycled within the farm—and not exported.) Pastured livestock manures, cover crops, perennial shrubs, bushes, and dead wood/trees. Meadow biomass, forest materials, and leaves, virtually all plant materials can and should be recycled on a balanced farm. Creating a landscape ecology based on biodiversity within the farm property promotes a farm's health and life—especially soil life.

Principle Three: The Farmer

The farmer is perhaps the most important factor contributing to and determining the success of a biodynamic farm system. A conscious, biodynamic farmer holds and implements the farm vision, in service with nature, connecting the land to its natural systems, and supporting the natural thriving of life on the farm.

What exactly must a biodynamic farmer do? The biodynamic farmer implements the principles of biodynamic agriculture daily. He, she, or they grow soil. He, she or, they align their will and heart forces with the heart or being of the property, manifesting its movement forward a healthy, productive future. They hold charge of the vision and the creative intention—the creative principal—for the land and its creatures.

Agriculture is a profession that deserves respect—even reverence—and gratitude. In the modern world, it would seem, that farming is a profession that many young people would not choose, even avoid, because of its perceived low social status. Plus, the long hours and hard, manual labor required, combined with (often) marginal financial rewards for many startup years—can be discouraging to young potential farmers. Yet, more and more youth, both young women and men are choosing to become regenerative organic-biological and biodynamic farmers. Why are these conscious individuals of the next generation committing to ecology-based agriculture?

In our rapidly changing and uncertain times, more young people wish to take back control of their lives. To move back to the land. To choose self-reliance. To make a difference. They have an increasing awareness that our world needs renewed cooperation with nature and that we as individuals must connect with the essence and substance of life. And, because the conscious farmer can make perhaps the greatest positive impact on climate change of any profession today—and the upcoming generation is the one waking up in greatest numbers to the climate apocalypse we are collectively facing down.

Principle Four: A Holistic System

Goethean science promotes the practice of seeing and learning consciously, from observation to perception. This perspective is how a biodynamic farmer views or looks at everything from the *plant*, and how it grows and feeds, to the complete farm system. A *holistic* system. The plant, the soil, and the individual livestock creatures are each living beings, not just component parts of a factory or a machine. Put another way, life and living creatures are not expendable products to be consumed, exploited, manipulated, and used up —all life is precious. A holistic system focuses on relationships. Biodynamic farm practices—e.g., the recycling of farm property resources, crop rotation, plant diversity, and the incorporation of livestock—create and nurture relationships within nature, making it possible to generate both quality life-giving produce and eventual consumer health.

This wise, ancient philosophy is intrinsically at odds with the dogma of industrial agriculture. Life comes from life, not chemistry alone! It holds an etheric body or force. Life can be built via the redemption of dead matter that previously had a life—that is, we can *regenerate*. But life cannot come from completely dead non-living substances—we cannot create life where there was never life to begin with.

This principle, profound worldview is fundamental to a holistic understanding of nature and all living systems—that is, the work of conscious biodynamics. Conversely, all synthetic materials of petro-chemical origins (tar, plastics, microplastics, motor oils, and fuel) are *dead* substances devoid of a life force. It must be understood that they will always carry the death principle wherever they infiltrate a farm and nature.

Principle Five: Cooperating with Nature and Nature Intelligences

Many wise, indigenous cultures have known for millennia that the earth is a living being. The earth breathes in and out with the seasons and in the daily rhythms. Soil breathes or respirates. It evolves. It is very much alive.

A biodynamic farmer is wise to set aside or allow for wild, natural environments to exist within the farm landscape—for the whole of nature and its influences bring great benefit to the farm and its prod-ucts. And, at the same time, if nature is an ecosystem and a farm is a portion of nature, a biodynamic farmer can (and must) be an active participant in that ecosystem, in a way that honors the natural envi-ronment. Specifically, while allowing natural environments to exist, an aware, active farmer assists with the cultivation of plants on natural land, supported by the subtle influences of the zodiac and moon rhythms (plant tides). A good, conscious farmer understands weeds—both beneficial and invasive species—and how to manage them. And, in response, a good, conscious farmer tends the earth with natural products: herbs support a delicate balance of health on a farm, especially concerning managing insect populations; biody-namic sprays enhance life and life processes; most importantly, the biodynamic preparations are gifts that a biodynamic farmer gives to nature and the land. Aligning with nature empowers the farmer to achieve his/her goals.

Awareness of plant and nature intelligence is a conscious farmer's key to opening the door to wider natural realities. When you act in service of nature, in a way that respects and cooperates with her intelligence, she gives her gifts freely and with love.

Principle Six: Growing Food for Nutrition

Food of true quality can feed not only the material, and physical body—it can also feed the human spirit. Foods produced by industrial agriculture are from the death culture and science. Materialistic agriculture processes are commodities produced for volume by weight. Biodynamic food grown for flavor and nutritional qualities, affords an opportunity, for human physical and spiritual development, providing for the needs of the human consciousness. Thus, quality, nutritious food grown consciously—with the dual intention of producing substance *and* respecting nature—may be characterized as spirit in matter. Consider, for a moment, how remarkable and powerful that is.

Principle Seven: Creating Healthy Social Connections

Respecting and acknowledging biodynamic farm workers and their invaluable contributions to agriculture is essential. And, taking that appreciation one step further, these workers thrive by respecting and interacting with each other. A successful biodynamic farm is a community-supported farm. Establishing a wider community and collaborating with neighboring, peer farm workers—supporting partnerships that strive toward common goals—are important steps in the process of building a successful agricultural business. Even family members working on small farmsteads must cooperate and find harmony working together.

Specifically, there are numerous ecosystem benefits of connecting with your biodynamic-organic community at a local and regional level—a local community of eco-farmers working together to produce quality food and educate the public. Each farm becomes more productive. Local appreciation for the value of biodynamic produce increases, a highly effective marketing strategy. Consumer trust in their food sources builds. In short, considering the wider community of life forms impacted by biodynamic systems—growers, end consumers, local chefs, local food distributors, and so many more—it becomes another clear example that nature and her intelligence are not to be underestimated.

The Australian Demeter Biodynamic Method!

With this foundational understanding of each of biodynamics' core principles, we can begin to understand what sets this method of biodynamic practice apart from the other comparable organic methods that have developed around the world.

The Australian Demeter Biodynamic (ADBM) method is based on Australian research and practice, primarily developed, over the last 60 years, by Alex Podolinsky and numerous practical pioneering Australian farmers.

Originally, this biodynamic method was developed to suit Australia's arid conditions, in particular. A livestock shortage (due to the vast spaces and prolonged dry conditions in much of the country) meant that manures, and thus compost, were often in very short supply. This posed a real challenge to applying typical compost preparations and led to the creation of a "prepared 500" horn humus natural alternative. Biodynamic compost preparations (502-507) were added after the 500-horn-humus material had been harvested from the horns.

The development of the full suite of preparations followed, with a process of experimentation and testing. This included the method of activating the horn 500 and 501 preparations exclusively in copper stirring machines (developed in Australia) and applying or spraying the preparations at precise times and under the best conditions. Then came quality testing of the biodynamic preparations and validation of results demonstrated by their application and use.

The ADBM is meant to address the conditions of our changing times: on a plant level, the breaking down of natural plant growth; on a farm level, the stressed conditions caused by forced farming methods; on a massive global level, the deteriorating environmental-climatic conditions worldwide. It stands to reason that these conditions call for a new impulse in agriculture practice.

The practical application of the ADBM is designed to:

- Develop and maintain soil structure.
- Develop the humus on which that soil depends for its fertility and resilience.
- Support the growth of healthy plants under the sun's influence.
- Grow produce of the highest quality in both flavor and nutrition.

Under Podolinsky's guidance—building off the insights of Dr. Rudolf Steiner and the scientific experimentation of Ehrenfried Pfeiffer—the biodynamic preparation quality was refined. The success of the method depends on plants feeding themselves according to the inherent processes of nature. Through nutrients that are naturally occurring and held within the humus content of the soil, not through artificially added, water-soluble NPK nutrients in the soil water. Beyond the application of preparations, the ADBM uses deep ripping or subsoiling to properly begin the soil structuring and soil development. Ideally, each biodynamic farm operates as a closed unit, with the farmer working with the land by using and adapting the preparations to naturally occurring, seasonal variations and fluctuations of the soil and climate.

When I first met with Alex Podolinsky at his home and farm in 1994, I was surprised to hear I was the first American farmer who had come to Australia to visit and learn from him. He gives freely of his time, advice, experience, and knowledge—without a fee—but he gives it selectively. So, I asked him what factors and criteria were most important to him when, making the decision, on whether or not to work with a farmer. Alex responded that he was looking for "a good man", by which he meant a good, moral human being. (Importantly, this was not a statement on gender. The practical reality was that, in his time, almost all working farmers in Australia were men.) Other criteria Alex reported looking for? Beyond being a moral person, Alex looked for partners who were (at least) third-generation farmers who owned their own farmland. Many people felt that Podolinsky was too restrictive or narrowly focused, but he was single-handedly responsible for Australia's wide acceptance and practice of biodynamic methods on hundreds of farms. (Of course, he would say that the dedication and ingenuity of the Australian biodynamic farmers had a very significant influence on the success of biodynamic farming in Australia as well.)

Primary aspects of the ADBM are quality-made biodynamic preparations, their proper storage, activation, and, subsequently, their appropriate and timely application. In Australia, the preparations are made on a very large scale—a scale matched nowhere else on Earth. (In one single Australian location, 125,000 cow horns are buried each year.) In Australia, I was taken by Podolinsky to observe the Australian scale of preparation 500: a large shed full of 55-gallon barrels full of horn humus (prepared 500); many hundreds of pounds of this dark brown, moist colloidal humus material.) Twice each year in Australia, thousands of hectares (on 800 certified biodynamic farms and many more gardens) are sprayed with prepared 500 horn humus.

Today—thanks in large part to Podolinsky's tireless and dedicated pioneering efforts—getting started with biodynamics in Australia is a simple process. As simple as filling out an inquiry form and enrolling in an Introductory Field Day that teaches and demonstrates the ADBD method practices, including:

- History and development of biodynamic farming and growing.
- Plant feeding within nature.
- Soil cultivation to improve soil structure.
- Soil humus formation and microbial activity.

- Biodynamic plant expression.
- Preparations 500-507.
- Using biodynamic preparation 500.
- Using biodynamic preparation 501.
- Stirring and applying the biodynamic preparations.
- Preparing the soil to receive the preparations at the most beneficial time.
- Building a biodynamic compost heap (using compost preparations 502-507)
- Sheet composting and green manures using Pfeiffer Field Spray.
- Using the biodynamic astrological sowing chart/calendar.

Field Day events are held throughout the year in various locations in Australia.

In America, we would benefit greatly from a similarly practical and uniform approach to biodynamic education and the use of biodynamic preparations. Across America and Europe, we find a wide variety of approaches to the biodynamic method and thus some inconsistencies and quality variations across individually developed approaches. Properly prepared preparations should be activated in a good stirring vessel, ideally made of copper or glass. Copper is the most practical, but clay, ceramic, and stainless steel are also used and are preferred over plastic or wooded vessels—as copper and glass are conductive materials.

As the foundation of biodynamic practice, let's dive into an exploration of each preparation individually.

The Bio-dynamic Preparations

The biodynamic preparations are each unique combinations of substances. And, each has unique requirements—related to how they are made, stored, activated, and applied—which must be intimately understood and expertly implemented.

Biodynamic farming involving preparations represents a very

different approach from organic farming, where applying mineral inputs in hundreds and thousands of pounds per acre regularly is as common as it is in conventional, chemistry-based agriculture. The materials and processes of these preparations do not involve material scientific thinking at all. While some people will understand these materials as homeopathic in nature, a more advanced understanding would be found in physics and quantum physics.

Dr. Steiner indicated that the preparations involved working with "forces," not material substances. Forces can be understood in the context of the discipline of physics. Dr. Steiner was ahead of his time, understanding that all matter has energy behind it. (Matter and energy are now known to be synonymous.) The biodynamic preparations work as forces—that is, these specially prepared humus materials carry forces and "information" when activated in pure water, allowing that water to then carry the forces forward and inform plants, soil, and the intelligence of nature. In other words, as a result of the application of these biodynamic preparations, soil life, and plant life are made aware or informed of the direction to develop. Information that is carried by the properly made, prepared, and applied biodynamic preparations.

I. 500 Horn-Humus

The horn humus begins as horn manure. Then, it is buried for six months—from Fall Equinox to Spring Equinox—and dug up and emptied. It must be stored in a cool, dry, dark location—in a storage box, barrel, and crock or glass—surrounded by sphagnum moss, if possible. It must not dry out or be subjected to excess humidity that might grow mold (this would be a rare occurrence).

While it *begins* as manure, horn humus is not manure! When made properly, the original green cow manure transforms or converts within the horn during the cold winter months into a total humus material. To apply the preparation, a portion of this horn humus is added to 10-25 gallons of warmed rainwater pure (without chlorine or other harmful chemicals) and then heated, using gas or wood-fired heat (no electrical heating is permitted near the preparations) to 100 degrees Fahrenheit, or 38 degrees Celsius. In order, to be activated, the mixture is vigorously stirred for one hour—no less—from vortex to chaos, back and forth, either by hand or using a copper stirring machine. When stirring, a 12-second interval between vortex and chaos is found to be ideal. 500 horn-humus should be moist, dark brown, sweet or neutral smelling, and colloidal.

500 is applied at a scale of 2 ½ ounces per acre—approximately the size of a large golf ball. It is applied on or near the full moon, from dusk to midnight, when the ground is wet or moist, either from recent rains or watering with an irrigation system. Large droplets of stirred 500 are sprayed (not a fine mist). Spraying or applying in windy conditions could be challenging and is usually avoided. And no rain or irrigation is to be received for 24 hours after spraying the preparations, so as not to wash away the effect of the spray.

II. 501 Horn-Silica

The beginning ingredient of 501 Horn-Silica is good-quality quartz crystal. It is powdered super-finely (talc-like) and water is added to create a slurry, which is poured into a cow's horns and then buried in the soil, in a sunny location. It is buried for the summer months, from the Spring Equinox to the Fall Equinox.

After it is harvested in the fall, the silica is powdered and then stored dry, in a glass jar in a sunlit window.

501 is to be used in very small amounts—4 grams, or less than a teaspoon. To be made, 501 is stirred for no less than an hour in rainwater at 100 degrees F, in the same manner as 500 horn-humus material. It is sprayed in the early morning—in conditions where humidity or moisture is present, in still air—in a very fine mist onto new leaf growth (before or after the flowering of plants).

501 is a fine foliar spray that assists the light metabolism of the plant. It can be used to increase the sweetness and sugar content of fruits. In addition, in conditions of excess rain or saturated wet soils,

it can assist in disease prevention by drying out the excess water, creating a crystalline effect with the plant and its roots.

III. 502-507 Compost Preparations

Each of the 502-507 preparations begins as a compost herbal material; 502 begins as yarrow flowers; 503 begins as chamomile flowers; 504 begins as nettle leaf; 505 begins as oak bark; 506 begins as dandelion flowers; 507 begins as valerian flowers (this core material is juiced and "fermented"), while 502 to 506 are buried in the earth (all but nettle leaves are placed in special animal sheaths). The compost preparations are applied first, before the 500 horn-humus and the 501 horn-silica. They are intended to balance the imbalances in soil minerals and correct mineral deficiencies over time. Each is made in complex processes requiring a year. Annual applications on the farm create overall health.

IV. 508 Horsetail grass (Equestrian Arvense)

508 originates from an ancient plant that has the highest known natural content of silica. And the bio-silica of this plant is unique and has different properties from minerals, rocks, and crystal silica (501). Fermented horsetail tea is recommended as a further potentizing of this preparation. Horsetail tea is sired for 20 minutes in 100-degree water (rainwater is ideal). It is sprayed on the plant and soil and is used to remedy and balance an excess of water and/or fungal issues. Repeated sprays are encouraged and helpful in greenhouse growing, plant or animal diseases involved with excess water, and on wet (ground) property.

V. Barrel Compound (BC)

Barrel Compound is a concentrated, more refined compost-like material made from fresh cow manure, basalt rock powder, finely ground eggshells (chicken shells are most common), and compost preparations 502-507. Two and a half ounces of BC are stirred for 20 minutes in 100-degree water and sprayed on one acre or less of land. This preparation has a short-term effect of about two months as it acts like compost and can be repeated as needed in intensive cultivation operations. This is a simple remedy to increase soil fertility and apply compost preparations. It is relatively simple to make and can be easily demonstrated to beginners interested in biodynamic work.

Sourcing biodynamic preparations can start with the Josephine Porter Institute JPI (*https://jpibiodynamics.org/*) in the US and local preparation-making groups can be found on well-established biodynamic farms. In Australia, contact *https://demeterbiodynamic.com.au/* Every country or region will have a website and access to biodynamic preparations or you can contact Demeter International for more information *https://demeter.net/*

6

THE SACRED COW

THE VITAL ROLE OF LIVESTOCK IN AGRICULTURE

A HARMONIOUS FUTURE between food and agriculture has a lot to do with the vital role that livestock plays in the agricultural system. Society at large is, thankfully, growing more and more aware of the currently poor relationship we have created with our farm animals as a result of industrial *modern* animal husbandry practices—and the consequences of that damage. There is a serious need to return to a traditional relationship with our farm animals. So, what is the balance we must create to establish and maintain a compassionate, healthy, and humane relationship with the cattle and livestock involved in food production activity? Is there a way to feed human populations and maintain healthy agriculture with or without livestock?

Historical Context

Both wild and domesticated animals enhance the natural landscape, the feeling of the environment, and the fertility of its soil. For more than ten thousand years, agriculture has been dominated by traditional animal husbandry practices involving the rearing of sheep, goats, hogs, and then cattle. We could include yaks, water buffalo, lamas, donkeys, horses, and birds "fowl".

Today's European domesticated cattle (Bos taurus) descend from the first cattle domesticated from the wild Aurochs of the Taurus Mountains (modern-day Turkey and Syria) approximately ten thousand years ago. To the conscious farmer, the Bos taurus—along with the zebu cattle of India and North Africa (Bos indicus), the water buffalo of South Asia (Bubalus bubalis), and yak of the great Himalayan plateau (Bos grunniens)—have been essential contributors to agriculture and civilization.

As are all our mammal cousins, the cow is a sentient, feeling being. Bovines (the cow, yak, and water buffalo) are revered and respected by many cultures for the indispensable role they play in agriculture. From the very beginning, in the earliest civilizations of the Indus Valley, Crete, Greece, Egypt, and Mesopotamia, domesticated cattle have been traditionally and literally held as "sacred." Sacred means *connected with God* (or the gods) or *dedicated to a religious purpose and so deserving veneration.* Cattle in particular are seen as sacred by Hindu culture, and by many biodynamic farmers as well.

Zoroaster and his followers say this about the cow. Cows are considered beneficent animals because they plow fields and produce milk. Killing a cow is forbidden because it is considered a great service to humans. The term geush urva means "the spirit of the cow" and is interpreted as the Soul of the Earth.

Livestock and the Land

Cattle play many roles in conscious agricultural systems. They contribute to and stimulate soil fertility through both their prized manure and their migration and grazing practices; they provide essential secondary nutrition (dairy); they are used as draft animals—and so much more. While cows have recently received undeserved bad press from vegans, environmentalists, and animal welfare enthusiasts, it is the mismanagement of cattle by humans—not the cattle themselves—that deserves justified criticism. (It is on factory farms, or CAFOs, where cattle are not grazed and live in filthy feedlot operations, that the reputation of cattle has been so sadly fouled.) On well-managed farms and ranches, cows are not a liability but a net agricultural asset, serving a vital role in supporting and building healthy ecosystems. Rodale Institute FTS shows clearly that the organic manure system of fertility building is the most profitable cultivation system, even without organic crop premiums.

Cows are one of the farmer's greatest allies (along with the earthworm and honeybee). Primarily, their grazing stimulates plant and grass growth, and their manure stimulates the symphony of the soil. When their contributions are well-managed, cows are soil fertility builders drastically helping to increase dung beetle and earthworm populations. Earthworm castings are the ultimate key to greater fertility.

Cattle and all domestic livestock are also the best managers of a farm and grazing lands. When a pasture is divided up into many paddocks—at least five on a small farm, or up to 35 or 40 on a dairy farm—livestock can be rotated regularly through them in a practice called Managed Rotational Grazing. This means that, after a pasture is eaten down, plants have time to grow back to an optimum height before being grazed again. After each grazing, many plant roots die and are devoured by the soil life. Plant roots reoccupy worm tunnels and, under a rotational grazing model, have the necessary time to grow progressively deeper, going hand in hand with soil structuring and aeration.

After grazing by cattle or other ruminants, pastures can also be harrowed to spread the valuable manure—a form of "sheet composting." The thinly spread manure is quickly digested and incorporated into soil humus by worms and microbes. (Harrowing is not necessary if dung beetles are active, and should be avoided when soils are wet, risking compaction by tractor tires.) There is no doubt permanent pasture is the best use of land for soil improvement in terms of biological activity, soil structure, and depth of soil humus development.

Zimbabwean biologist and livestock farmer Allan Savory has

famously observed that a single cow on a property for one year generates substantial, measurable benefits. A cow produces 29.5 kg (65 pounds) of feces or manure daily which is 12 tons (10,765 kg) a year. in a year[i]. Her manure or dung is made of up to one-quarter microbes which contribute substantially to the inoculation, regeneration, and replenishment of soil biology.

Today's Imbalance

If all of this sounds so straightforwardly advantageous, what is the problem? The problem, of course, comes down to the abusive and ruinous industrial agricultural practice. Overgrazing and poor livestock management of all too many farm properties have shown how the mishandling of livestock cattle, horses, and sheep can ruin a property. Essentially, industrial agriculture has created unhealthy environmental disasters involving livestock, disasters which are responsible for many people choosing to become vegan—not the livestock themselves.

As we know, carbon accumulation in our atmosphere from soil destruction, permafrost melting, burning of forests, and fossil fuel misuse is our environment's biggest existential threat—and methane is a major contributor to that. There are those false critics out there who try to blame cattle for methane production! But the *overwhelming majority* of our methane problem is not coming from methane from cows. In a healthy world, methane gas naturally excreted by the cow's rumination is in balance. Methane *pollution* comes from unnatural fossil fuel production and extraction—the wasteful infrastructure of natural gas, and its inefficient burning. This far exceeds any methane threat produced by cattle. It must be said that the net benefit of cattle and ruminants far outweighs the negatives of the limited cattle methane produced when seen in the light of good farm management. Soil microbes in healthy soil benefit from manures and are believed to mitigate methane produced by cattle

---

[i]   Extension Utah State University: *How Much Manure will my Animals produce?*

and other ruminants, thus even neutralizing or eliminating any cattle-attributable methane concerns.

Cow Horns and the Conscious Farmer

Cows and their horns are a subject that most people—and even most farmers—have given little thought to. Horns are a natural expression of cattle. (Unfortunately, in recent times, this truth has been impacted as dehorning and breeding of polled or hornless cattle have become more common.)

Dr. Steiner had this to say in his Agricultural Lecture:

Have you ever wondered why cows have horns? It is a most important question. I have just mentioned that everything organic, everything living, does not always need to have streams of forces that are only directed outwards but also streams of forces flowing inwards...What happens in the places where hooves and horns are growing? A locality is created where the inflowing stream of forces is particularly strong. The outside is strongly closed off. No longer is there any communication through permeable skin or hair, the gateway for that which seeks to stream outwards is completely closed. The formation of horns is in this way intimately connected with the whole form of the animal...The cow has horns so that it can send into itself the formative astral-etheric principle, that it can penetrate and stream into the digestive organism. What radiates from the horns and hooves can then work strongly into the metabolic processes. (Rudolph Steiner, Agricultural Course, lecture 4)

In biodynamics, the cow horn is essential for making the 500 horn-humus and 501 horn-silica preparations. As such, horned cattle herds and their propagation are critical to a healthy and fertile agricultural future.

Biodynamic Pastures

In his book *Back from the Brink,* Australian farmer Peter Andrews describes a visit to one of the top racehorse stables in England in the 1960s. When he complimented the stable manager on his pastures, saying that they appeared to be free of weeds, the manager became visibly upset. Peter later discovered that traditional English horse breeders considered that a good horse breeding pasture should contain at least eighty weed species. In fact, they held that if the pasture had less than forty weed species, it was in decline. (Not to fear—when Peter had a closer look at the pasture, he actually found many species of weeds growing, in large numbers.) These breeders also considered it essential that pastures should never be plowed, as their experience had shown that plowing a pasture meant it was no good for young horses for five years and would not produce a Group 1 winner for ten years—due to the loss of soil biology and the time necessary to recover from indiscriminate plowing. Commensurate practices and systems work well on a mixed farm, incorporating techniques such as open field/meadow pasture cultivation and the intergrading of livestock grazing with growing row crops and cereal grains.

Biodynamic pastures are composed of many plant species. The more the better! (That is, the more life and greater diversity the better.) Specifically, 60-90 species are ideal. Each plant species—each with its own unique biological activity, with its own community of fungi and bacteria, enzymes, and coenzymes—brings a different quality to the soil and specific benefits to the animals that graze. The large and varied combination of species gives the farm's soil food web a sufficiently varied palette to work with, to fulfill its aim of developing stabilized colloidal humus and bringing a balanced supply of nutrients to plants.

Similar to Andrews' findings, limited, well-managed weed populations are generally regarded as useful contributors to the overall nutrition and biological activity of a biodynamic pasture as well. In fact, many so-called "weeds" that would be eradicated on conventional farms are seen as valuable to the conscious biodynamic farmer in improving the digestive disturbances of regenerative-organic farm livestock. Native grasses are similarly encouraged on diverse farms.

From Livestock Nutrition to Human Nutrition

The error and great tragedy in agriculture is that most of our livestock, poultry, pigs, cattle, and farm ruminants are fed enormous amounts of grain. Ruminant animals are very efficient digesters of plant fiber or grasses, with their four specialized stomachs or digestive organs. But they digest grain with great difficulty and in a very environmentally costly way. And still, half of all farmed agricultural lands and water resources produce grain for livestock. This mismanagement of agriculture's grain treasure trove is untenable. It results in a great burden on our food systems, the environment, and human health. Vegans and vegetarians' please listen up!

The truth is that healthy livestock promotes human health as well, by virtue, of healthier food that livestock produce or help cultivate. Human nutrition is vastly improved when consuming ruminants' dairy products and/or secondary food products. Specifically, many plant nutrients are not easily, directly available to humans—or not available at all—through human digestion. We can only access the important nutrition from many fibrous plants after it has been converted, by ruminants, into the form of dairy products and meat. This fact was critical to early agriculture in northern climates where crops were not adapted nor often successful. Dairy products offered a critical food source in human development for many early human civilizations. This is our heritage.

Biodynamic and Regenerative organic practices and pastures promote healthy ruminant practices that promote human health in turn. For example, well-managed farms allow twelve inches of standing meadow or pasture to develop, leading to deeper roots, improved soil water retention, enhanced plant growth, and more. Photosynthesis sequesters carbon from the atmosphere in plants and their roots in the form of chlorophyll, polysaccharides, carbohydrates, sugars, and cellulose. Carbon returns to the soil when plants grow, die, and are eaten by ruminants returning as manure. Legumes, in partnership with their co-working rhizobia bacteria, bring atmospheric nitrogen into plants' roots and into the soil. Various herbs and "weeds" concentrate rare or essential elements—for instance, when copper becomes scarce, heliotrope will regenerate this element.

Under current chemical-driven agricultural systems, most farmers believe they must export large amounts of produce to make a living. This stands in contrast to traditional farming practices. Just 150 years ago, much of a farm's production was recycled within the farm system as livestock feed, and then animal and plant waste materials were used efficiently. In these generations' past, much of the farm's production was retained to feed large farm families and their workforce. This was a truly sustainable system in which only limited amounts of outside products were exported from the farm. (Importantly, changing climates further today's move towards these advantageous, self-sustaining systems. Temperate climates are much gentler and more conducive to no-input biological farming. Today, arid lands are being cultivated for extractive production agriculture. In Australia and the American southwest, long drought periods depress biological activity and lower organic matter and humus levels in the soil. As a result, in practice, Australian biodynamic farmers do generally have to import some non-water-soluble fertilizers from time to time, albeit at considerably lower levels, vastly less than conventional farmers.)

Today many conscious farmers are using a layered livestock farming system. Conscious agriculture will always incorporate cattle and domesticated livestock to improve and enhance soil and its cultivation. Livestock has always been integral to good agriculture. Can we return to practices that demonstrate appropriate compassion and respect for the sentient beings that we interact with so crucially in agriculture? Will we maintain and uplift these animals as one of the creations instructed by Zoroaster?

Looking to the Future

So, where do we go from here, with regards to interacting with our sacred livestock partners? Some emerging trends are undeniably ominous. Other scientific initiatives show promise.

It must not be overlooked that meat is a by-product of animal production. (Meat should not be the principal product of animal husbandry.) As meat could become scarce—another byproduct of industrial devastation—a particularly concerning recent development (that may continue spreading) is the production of "cultured meat" or meat tissue that is grown in laboratories. This is not science fiction: lab-grown animal tissue has already been developed; it is already being marketed to the public and sold. To be clear, this is another perversion of nature. Another example of ugly materialist thinking taking us closer to societal madness. Fake flesh is grown without a living, sentient animal body, heart, and blood. These artificially grown meats can never provide essential nutrition or warmth, not to mention the astral forces provided through natural livestock production. Dear God, we seem to be on the road to the world of the 1973 *Soylent Green*, when the food of a dying planet of the future was predicted. Let us not take this road to civilization's ruin and destruction. Life is sacred. Cows are sacred beings. This is old, wise knowledge. Cattle should be seen and respected, not perversely imitated in a laboratory for unhealthy mass consumption.

As I said, some emerging insights on the other end of today's scientific spectrum are encouraging. As a very useful bioresource for sustainable development, cow dung is widely studied. It hosts a wide variety of microorganisms varying in individual properties. (For example, cow dung microflora can contribute significantly to sustainable agriculture and a farm's energy requirements; it is one of the bioresources of this world that is available on a large scale and still not fully utilized.) An enhanced understanding of the mechanisms enabling cow dung microbes to degrade hydrocarbons is estimated to promote the bioremediation of environmental pollutants. With recent advances in scientific research and techniques for complete genome sequences, the genes responsible for bioremediation can be identified.

Another area of emerging study concerns developing microbial enzymes and antimicrobials. The production of enzymes by microorganisms from the cheap cow dung bioresource can find wide applications in various fields such as agriculture, chemistry, and biotechnology. The application of cow dung microflora with considerable antimicrobial potential can result in the promotion of human health. (However, comprehensive screening of these microorganisms to produce antibacterial, antifungal, and antiviral metabolites needs to be investigated.)

In India, dried cow dung is a major source of fuel for rural and poor populations. For millennia, cow dung mixed with plant fiber (straw/reeds) and soil has been used in house constriction, a process known as "cob." Dried cow dung plays another role in the Hindu purification ceremony and ritual known as Agnihotra, referring to a healing fire from the ancient science of Ayurveda. This practice is now being globally applied, assisting the healing and purification of farmlands and communities.

It is certainly evident that more detailed studies of the many potential features and applications of cow dung are needed, as—as funny as it might sound—cow dung offers us a great deal of hope. It holds tremendous potential for sustainable development as an easily available bioresource. I suppose you could say that today's science is beginning to understand what many cultures have known all along: that a lot of the answers come back to the sacred cow.

7

———

THE STRUGGLE

NATURE AND LIFE VERSUS THE CULTURE OF DEATH

WE HAVE REACHED A CRITICAL POINT, during human existence on Earth. As the age of materialism has magnified our continued loss of a spiritual perspective, we find ourselves delivered at a crossroads. In one direction: nature and life. In the other direction: the *literal* culture of death. Ultimately, this is the struggle between consciousness versus unconsciousness playing out with humanity. Our collective consciousness affects all life forms, including the Earth—also a living being. When humanity is unconscious, all humans are more easily deceived, manipulated, and controlled.

The Scientific and the Spiritual

At the core of this struggle are two opposing world views: the scientific and the spiritual. The atheistic scientific view is one that, has never agreed upon a definition of life. The best, loose definition upon which it can come close to agreeing is that life is a chemistry-based, Darwinian evolution system capable of self-sustaining replication. (This outlines a soulless, spiritless, random existence.) How empty and sad! The other, spiritual worldview—that of most cultures of history, and one on which most human beings agree—holds that there is an intelligently ordered intention to creation and life; that life is sacred and infused with spirit and soul. This is old, Indigenous wisdom, Celtic wisdom that has been largely lost—wiped from mainstream consciousness—at a great loss to the earth and humanity.

"Science without religion is lame, religion without science is blind"
    Dr. Albert Einstein

If we hope to survive as a species, we must choose the spiritual path. We need to rediscover, return to, and preserve this essential knowledge of nature and our connections with it.

Indigenous scholar Dr. Gregory Cajete, of the University of New Mexico, retired, has written several books, including <u>Native Science</u>, in which he provides education on indigenous cultural, ecological, and philosophical knowledge, which he aptly refers to as native science. He poses three fundamental questions, drawing on the ecological traditions of Indigenous peoples, about our modern dilemma:

1. How do we develop a better relationship with the Earth?
2. How do we develop a better relationship with each other?
3. How do we develop a better relationship with the entity that guides it all?

Cajete speaks about our relationships to place, environmental justice, interrelationships, and connectedness. How the native cultural view of the world is one of many worlds—a multiverse. (In fact, he addresses more about our universe of matter than does the construct of modern science!) The core perspective he teaches is that everything in nature is "alive" with energy (spirit). He states the goal of native science is to become open to the natural world with all of one's senses, body, mind, and spirit. This is the spiritual view at odds with the materialistic culture of death we have embraced.

Good Versus Evil—A Familiar Binary

The concepts of good and bad—life and death—exist all around us. As we've seen, good nutrition (life) in our food is determined by the soil, the plant, our oceans, and the health of animals involved in the food's creation. Continuing this line of thinking, our broader environment connects the larger *web of life*—i.e., the lower and higher life forms on the food chain.

Death is the flipside of this equation. In modern society, under modern industrial practice, infiltrating agents and substances of death are disrupting the cycle of life, soil life, water life, and our immunity at every turn. Death forces are effectively stopping the *spiritual stream* of living forces that are an essential part of human nutrition and consciousness.

There is reason some of this may sound familiar. In popular culture, this identifiable saga plays out frequently in movies—*Lord of the Rings*, *Avatar*, and *Star Wars*—and other media, when a world of *good* or *life, consciousness* is at odds for survival with darkness and evil that is bent on the destruction of consciousness and freedom at all costs. (Think the "Dark Side of the Force!") And in the far too real world, the ultimate battle between good and evil is seen in war. Wars are far too common in human history and are currently ongoing. They are the tragic epitome of the competition between life and consciousness, and death and destruction.

It Doesn't Have to Be Either-Or

In many ways, Dr. Steiner uniquely bridged this divide. While he was a scientist by education, he was also a person with exceptional inner sight, developed senses, awareness, and perception who knew it was essential to speak of the world as it truly "is." While it would have been politically correct, in both his time and in our present time, to remain silent and not address the elephant in the room—the *universe*—he had the power to see the world as it is. A complete whole with parallel ethers, dimensions, or the *realities* that we exist within. (This is a model that is now actually supported by quantum physics.)

Steiner acknowledged that the rise in prominence of scientific, and technological rationalism would necessitate, in the beginning, a corresponding reduction in more spiritual ways of knowing. But it does not have to be this way forever. While science isn't going anywhere, these two worldviews do not *have* to exist at such odds with each other. It is time to sort out the assumed conflict between a worldview that is *spiritual* and a worldview that is *scientific*, to not only look for the differences between the two philosophies but also find the bridges that connect them.

Humans are innately spiritual or non-material—as is all matter. Ask any scientist! Even the physical human form or body is but a scant two pounds of mineral earth and the rest of us the other 98% are other cosmic elements, or cosmic dust. The lighter elements that primarily make up our bodies and our physical planet—hydrogen, carbon, oxygen, nitrogen, and a few others—are cosmic substances that also make up cosmic dust, stars, and galaxies. We are all connected. The spiritual and religious systems that have developed throughout all cultures of the world over time have attempted to address this reality with varying kinds of wisdom. But many were developed long ago before we had the scientific understanding or knowledge that we possess today that provides rational arguments to support spiritual and religious beliefs.

Unfortunately, as a reaction to the brutal control that religious dogma has historically wielded across many Western cultures, modern intellectuals—the forerunners of modern science—began to reject a spiritual worldview outright. Out of that rejection arose a scientific view of world phenomena and matter that is theoretical and fractioned at best. This is what gave rise to today's materialistic intellectual class.

These modern-day, scientific intellectuals have deduced that God is dead or never existed. But the reality is that the cosmos—our universe—is ordered; it is *alive*; it is perhaps even conscious. The great expanse of existence is not random chaos; it is ordered chaos.

By all these accounts, it appears to have been created by some intelligent force or consciousness.

Let's talk specifically about the concept of ethers. Ethers are more commonly known as dimensions. (Think, "the other side.") One hundred years ago, almost all scientists rejected theories of ethers or other subtle dimensions. Yet, the term ether was used by Albert Einstein, who—as an exception among his scientist peers—acknowledged we exist with at least one other dimension that our senses normally do not perceive. (Today, ethers can now be measured. As such, they are now accepted as realities by the science of quantum physics.) Consider this one more example of many scientists taking a long time to catch up to the insight that most of humanity knows intuitively—particularly as our modern vocabulary evolves to better articulate these unseen realities. Intuition is old wisdom. It can inform us about facts and reality. This was observed and known by Carl Jung, the Swiss psychiatrist, psychotherapist, psychologist, and pioneering evolutionary theorist, founder of the school of analytical psychology. We can trust it when it tells us there is an invisible reality behind and parallel to the physical reality. Spirit worlds, fairy worlds, astral worlds, and heavens, perhaps? The mind expands with possibility.

In truth, there are many examples of science taking its time to catch up to inherently commonly known knowledge. The unseen atom was correctly described by the ancient Greeks. Today, millennia later, we can verify the reality of atoms with scientific testing. And, recently in England, an atom was photographed for the first time. Does the existence of that photograph make the atom more real than it was thousands of years ago? Of course not. But by making it more tangible, through the physical sense of sight, it becomes impossible for even doubters to deny.

On the heels of 200 years of intellectualism, the brightest minds in science today can now verify and prove that we live in both a non-material reality (or a *spiritual universe*) and a material physical universe—*and that this is not a contradiction.* Quantum physics proves that we live in a multi-dimensional universe with parallel realities—

that is, a multi-verse that our very limited five senses do not perceive. While these realities are, for most of human experience, the vast majority of people unseen, we interact with these dimensions daily. And, on occasion, throughout recent history, some exceptional individuals have come along who possess extra sensory perceptions of seeing, knowing, intuition, and inner sight, permitting them to see into these other worlds, dimensions, and the ether in some way. And their exceptional knowledge thereby uplifts the common knowledge of the masses. Dr. Steiner was one of the rare individuals, apparently who could experience more than the limitations of our typical five senses allow us. He stated that we have potentially twelve senses available to us, for those of us who are interested in them, open to them, and able to develop them.

The Culture of Death

I'll say it again: a spiritual worldview and a scientific worldview *do not have to exist at odds with one another* in a broad, theoretical sense. It is the damaging "science" of the industrial machine specifically that is at odds, today, with the ways of life and Nature.

The petrochemical industry has unleashed on life and nature an overwhelming deluge of toxic, raw, and processed petrochemical products of death—including plastics, forever chemicals-PFAS, pesticides, solvents-detergents, fuels, and micro-plastics. Since the atomic age and the predictably destructive global impact of radioactive materials, this destruction of life has only escalated. Perhaps even more disastrous is the daily stream of poison affecting our health in obvious ways—the forces of death—contributing to the cancer crisis, the astronomical recent rise in rates of immune disorders, Alzheimer's disease, Parkinson's disease, metabolic disruption, autism, and widespread poor nutrition and malnutrition.

One especially dangerous culprit is synthetic materials (man-made chemicals) in our food today. And without exception, *all synthetic materials and substances are dead,* their use is forbidden in all certified organic and biodynamic farms and foods. *They introduce death wherever and whenever they infiltrate nature and the farm.* All synthetic substances are prohibited on all Demeter-certified biodynamic farms. They are supposed to be prohibited by USDA organic certification as well but unfortunately—and perhaps predictably—exceptions are made. For example, diesel, and oils of all kinds are used in farm equipment on most organic farms and spills and leakage are inevitable.

Another glaring exception permitted in organic certification is the use of the synthetic amino acid methionine, which has been allowed for at least thirty years in all "certified organic" poultry-livestock feed. The best natural plant source for methionine (an amino acid essential to growth) is sesame seeds, which have not been produced as certified organic in the United States. (The reality is that all organic sesame seeds are hand-weeded, thus being very costly, and are being imported from Ethiopia or South America where labor is cheap.) The additional cost of food-sourced methionine is largely blamed as the limiting factor that has caused most organic livestock feed producers to avoid using organic sesame seeds in their feeds. But this excuse ignores the real, larger problem involved: a lack of vision and will within the farming community and USDA organic program to solve these kinds of issues creatively. They could support the development and incentivization of a mechanical cultivation solution for the domestic production of certified organic and commercial sesame seeds, but they don't.

Perpetuating the carbon myth that dead petrochemical carbon-based materials are organic—and keeping the wider public ignorant of the shortcuts being taken—is a great human health disservice. Often, the term *organic* as it is used in chemistry realms is confusing at best, and misleadingly wielded at worst. Carbon found in diamonds or slate is not alive, or organic—it is biologically dead. Carbon in asphalt and crude oil is not alive. Even carbon as a gas in the atmosphere is dead (inert) until it becomes a part of a living plant or organism. It's death—dead-carbon. It will not, *cannot*, support life or life processes. This is why synthetic pharmaceuticals; fifty percent of all drugs are so ineffective and have dangerous side-effects. On the other hand, carbon in soils (soil organic carbon; humus, humates, humic acid, and carbonic acids) and all carbon in wood, cellulose, plant lignin, sugars, polysaccharides, plant exudates, necromass, and green leaf/root carbohydrates are living carbon compounds, with a life force an etheric life force. Living organic carbon storage is made possible by other living or dead plant host materials.

The future of Medicine is plant-based medicine, Dr. Jeffery Bland, is a thought leader who has spent more than four decades focused on the improvement of human health. He is known worldwide as the founder of the functional medicine movement, which represents his vision for a care model that is grounded in systems biology and informed by research that he has a unique ability to synthesize. His pioneering work has created the Personalized Lifestyle Medicine Institute (PLMI) as well as The Institute for Functional Medicine (IFM).

What makes the difference between living carbon substances and non-living carbon substances? Their molecular form and the predominant electron spin. (Electrons are buzzing around and behaving as if they are twirling around on their axes like spinning tops. These "spinning" electrons are fundamental to quantum physics and play a central role in our understanding of atoms and molecules.) The form of carbon (and all elements) is measured by the predominant spinning direction of the molecular atom's electrons. Specifically, the predominant spin of the electrons is either descending (right spinning downward) or ascending (left spinning upward). This is referred to as *delineation*, and it is not new information. This fact is well-known by big science but ignored for convenience and profits when creating synthetic molecules (chemicals).

All living molecules of life and their atoms have electrons spinning in a predominantly ascending left direction—towards light or the sun. Conversely, non-living molecules' electrons spin in a predominantly right direction. Right-spinning delineation indicates that a molecule cannot support life. All dead or biologically finished substances (such as uric acid in urine, and petrochemical products derived from *non-living carbon* crude oil and asphalt) have their carbon (atoms) electrons spinning predominantly in a downward, right-spinning direction, or a right-spinning delineation. This is atomic and molecular physics, it's basic chemistry. And still, unfortunately, it is ignored by scientists and chemists who avoid taking moral responsibility for the mass production and proliferation of death forces in petrochemical products widely used by the consumer.

I'll put it plainly: all synthetic compounds are biologically dead. They carry death, to all living tissue and all living organisms. Synthetic amino acids, synthetic vitamins, and pharmaceutical drugs derived from petrochemicals are dead, *even when* they mimic or nearly mimic their natural, plant-based molecular counterparts. Whenever synthetic materials chemically interact with living matter, tissue, or organisms, they are ineffective or will damage, mutate, or kill them. How, exactly? All poisonous synthetic materials still interact with substances around them; in their journey to become disarmed and reach a balanced state with the world of nature, they go through a process (reaching equilibrium-balance) of breaking down. They damage and steal electrons from balanced, stable molecules part of the living world. In other words, these substances act as *molecular terrorists*, taking out or killing living molecules and organisms in great numbers before ultimately becoming neutralized or disarmed themselves.

These synthetic materials of death could rightly be characterized as immoral creations—products of the evil unconscious of mankind. Take petroleum as a catastrophic example. Crude oil and petrochemicals are derived from matter that should have never been taken out from under the earth. Petroleum is called black gold when in reality it is the black death. It is the greed of energy demands, and consumption used to drive the unnatural *growth* of all global economies. Unfortunately, this crude oil upon which our civilization has become so highly dependent should never have ever been dragged from the bowels of the earth's crust and brought into the sphere of the living planet and the light. For anyone who cares about impeding and stopping the forces of death and destruction toward nature and life, these are basic table stakes. It couldn't be any simpler: *stop using synthetic substances, and products whenever possible!* Stop buying conventional-grown industrial-chemical food products!

Of course, the trend of death does not stop there. AI energy needs are promoting the revival of further nuclear energy use. This death principle will cost humanity and our living planet further if developed. The assault on life takes on a whole new level with the proliferation of two horrifying trends—GMOs and cloning—which have both once more upped the ante and elevated the risk of the natural order's complete disruption. They represent a kind of de-evolution of living organisms and even humankind.

A sad reality of modern-day biology and science, in general, is that there is no understanding of life and what it is. There is no consensus on the definition of life. What is the spark of life the animator or life? There are numerous current definitions accepted for life, but all are inadequate. Here is one simple concise definition: *Life is a self-replicating chemical system that uses energy to keep itself from reaching chemical equilibrium.* This definition is wholly inadequate—it completely ignores and invalidates all spiritual wisdom of the ethereal spark that animates life (matter-spirit manifested). Call it what you want: spirit; prana; mana; inner essence or animator of life. There is a dangerous kind of purpose to the exclusion of spirit (the sound, bani-sound, vibration, frequency). Thus, all manner of manipulation and destruction of life is acceptable and tolerable when life is not sacred and has no spirit or soul.

We must ask ourselves: why has death taken hold of the collective human consciousness? Is it the fallen angel saga of Lucifer and Ahriman playing out their drama? Is it the weakness of human temptation? Is it the end of the Kali Yuga and the fall of humanity once again? Is it the struggle between the Dark Forces and the Forces of Light? The Force may be with you, or it may not, and this seems to be our ultimate choice to make at this turning point in human history.

Each person will make their own decisions and pass their own judgment, but one truth is certain: as a species, a great climax is upon us, upon all life forms, and upon the Living Earth, which is home to us all. Humankind is a force of nature, clearly, but what kind of force will we ultimately be? Are we to be a cancerous growth bent on destroying our mother? Our own life source? The Living Blue Planet? Or are we evolving human beings willing and able to consciously right our course, for the betterment of all life forms? If we choose the right, latter path, how will we succeed? How do we stop the boulders of death rolling already down the hill? How do we save ourselves and our natural home? It will require the evolution of the collective consciousness of humanity to a position that once more chooses life and consciousness. And it will require a great many of us. Our collective conscious participation is necessary to reverse our fate, and time is running short. If humanity fails to act soon to make major changes favoring life and nature, the impending backlash will cost humankind dearly.

Waging a War on Nature

I want to highlight one more area of our natural world facing death and destruction at our own hands: destroying Wild Nature, including deforestation. Clearing the forests by the method of burning is a practice that continues today in the Amazon and Borneo rainforests. This is a global travesty. Add the global increase in wildfires ravaging millions of acres per year in Europe and North America. Oftentimes burning forests is done to turn these forests into agricultural properties; specifically, we are currently destroying 18,000,000 acres of forest per year to create additional pasture and croplands. What a dangerous, backward practice. This activity is eroding the soil. It is destroying critical ecosystems and decimating many irreplicable species. And—the ultimate irony—it is being done today primarily to increase completely irrational and unsustainable capacity for synthetic production: cheap fast-food burgers, mono-cropping of palm oil plantations, and more GMO soy fields.

The Amazon Forest ecosystem has lost 17% of its original glory, causing extreme drought events in recent years. It has been predicted that a 25% loss will be the tipping point to the destruction of most of the Amazon's tropical rainforests, but also a global tipping point that very well could cause a collapse of weather systems and established weather patterns all over the world. This is very serious and looming large!

Consider this harrowing figure: It is estimated that, before human civilization, the Earth was estimated to have 5.8 trillion trees. Now, Earth is left with only about half—2.9 trillion trees. At the present rate of forest destruction, how much of the world's forest will remain? This is the struggle between nature/life and the death culture epitomized and played out before our eyes.

Waging war on nature will always be a fundamentally losing proposition. How can this not be abundantly clear, to everyone? As a civilization, *we cannot continue the assault on the very nature and life that sustains us*, but the rate of continued destruction is staggering. A continuous poisonous stream keeps flowing through our automobiles, farms, and equipment (construction, transportation, and shipping lines). Through our homes, our factories, and our land. Through our rivers and oceans. The poisonous death culture of our industrial age is everywhere in our civilization, coming from the very top: the world's largest 90 corporations produce, by far, most of the greenhouse gases.

The stakes could not be higher or more real. There is a very real possibility of millions of life forms soon going extinct. *What is being referred to as the sixth mass extinction is upon us.* The Holocene or Anthropocene extinction is an ongoing extinction event of species during the present Holocene epoch, *caused by human activity*. If we continue along our present trajectory, scientists are predicting that within 20-100 years, half of all species of life currently on Earth will be dead and lost.

We are experiencing an assault on all life. A biological annihilation. Dr. Paul Ehrlich, professor of biology and population studies at Stanford University, said that "when humanity exterminates populations and species of other creatures, it is sawing off the limb on which it is sitting, destroying working parts of our own life-support system... There is only one overall solution, and that is to reduce the scale of the human enterprise...Population growth and increasing consumption among the rich is driving it." Ehrlich also points out that while habitat destruction and deforestation for agriculture and pollution are the primary culprits, climate change dramatically further exacerbates both problems, which could have disastrous consequences for the ability of many species to survive on Earth.

Our Current Crossroads

There is a Persian proverb that says: "Whoever creates a garden becomes an ally of Light; no garden having ever emerged from the shadows." And Sir David Attenborough says this: "The truth is: The natural world is changing. It is the most precious thing we have, and we need to defend it."

All hope is not lost. We have a path forward. The positive and constructive alternatives to our present destructive petrochemical and radioactive culture of death exist and are at hand. They all come back to agriculture and consumer choice. How we treat and interact with the natural world and environments. It is, the choices available to us about the foods we purchase, be they biodynamic, organic, or chemically, conventionally produced. As consumers, our great, collective purchasing power is what can change the tide of our present culture of death. Will we think critically and arrive at objective truth? Will we recognize the agents of death? Will we choose life, nature, and our health? It should be very clear that we need nature to survive and thrive. Will we embrace life and consciousness?

8

———

PRACTICING CONSCIOUS AGRICULTURE

THE FUTURE OF FOOD AND NUTRITION

WITH EIGHT TO TEN billion people inhabiting the blue planet, restoring balance and harmony with nature will be no easy task—but it is possible, and it is necessary. Conscious agriculture is the solution, using a sophisticated combination of techniques: biodynamics, regenerative organics, permaculture, carbon farming, keyline design, agroecology, agroforestry, holistic management, and planned grazing. As the United Nations puts it:

> Needed are innovative systems that protect and enhance the natural resource base while increasing productivity. Needed is a transformative process toward holistic approaches, such as agroecology, agroforestry, climate-smart agriculture, and conservation agriculture, which also build upon indigenous and traditional knowledge. Technological improvements, along with drastic cuts in economy-wide and agricultural fossil fuel use, would help address climate change and the intensification of natural hazards, which affect all ecosystems and every aspect of human life. Greater international collaboration is needed to prevent emerging transboundary agriculture and food system threats, such as pests and diseases. *(Food and Agriculture Organization of the United Nations, FAO)*

In short, we must align with nature and assist nature to recover and rebuild healthy ecosystems that benefit eco-farming and humanity. Through mindful planning and actions, we can have both healthy environments and healthy balanced agriculture. The European Union (EU) just passed an ambitious new goal. This legislation by 2030 is to restore 20% of land in the EU to Nature. Nature restoration will reduce large industrial agriculture and encourage smaller diversified ecological agriculture.

It could be said that Earth's evolution—or its demise—is now in the hands of humankind. Intelligent people must make wise choices based on the wisdom of nature. Good science can be an ally, giving insights into the planet's natural, nontoxic, functional, and practical solutions to soil health and fertility. Consider the impact of every decision we make. Conscious farmers—those who care passionately about life quality, farm products, and how that product nourishes people and families—will play a great part in nature's restoration.

A few years ago, I was invited to visit one of Australia's oldest biodynamic farmers (and the first biodynamic rice farmer) and his wife on the Murray River, in the Murray-Darling Basin. I was treated to a simple but fabulous midday meal; the entirety of this meal came from their farm. This couple became biodynamic farmers, then in their seventies, because of their interest in producing and eating good nutritious food. The rice, beef, peas, peaches, and ice cream of this meal were lovingly produced, grown, and prepared by the two of them. The complete meal was natural and unprocessed food, rare in modern fast-paced societies. That is a real passion and commitment to conscious living and farming. We need much more of that, on a global scale.

In figuring out how to implement sophisticated, conscious strategies, we can learn from the knowledge and wisdom of those who lived in the Americas before Europeans arrived. Particularly relevant native Indigenous strategies include:

- Decentralizing human populations.

- Intentional land management, including wildlife habitat expansion.
- Anthropogenic influences that benefit nature, include controlled burning and grazing of the prairie landscape to expand and enrich grazing lands.
- Feeding the food web and planning for perpetuity.
- Using watersheds as productive agricultural land (an alluvial farming technique).
- Healing the soil.
- "Hozho"—embracing, with joy, our part in the earth's systems to protect and augment life. (From *Lyla June, Ph.D., and scholar.* Healing Nature through the Dine' (Navajo) view of life.)

As a people and culture, we must return to existing in harmony with Mother Earth. The flow of misinformation, corrupted information, disinformation, and conflicted opinions are obstacles used to create apathy, inaction, and stagnation. Tides will only begin to turn when we achieve a fully conscious, collective society; when public opinion is won; when good, bias-free education is the norm; when many more millions of conscious consumers begin to choose with their purchasing power, for the consumer voice is the one driving force with the power to insist that government and corporate policies change. In short, complacency and inaction will be overcome by motivated, passionate individuals involved in making change. Young people like Greta Thunberg, for example. That is, the generation, who is either going to reap the greatest benefit from societal change or pay the greatest price for maintaining the status quo of inaction.

Looking Back and Looking Forward: A Matter of Time

Our path forward is twofold: Farming practices that recycle essential nutrients, and land management practices that are conscious and balanced. In a broad sense, this is ecological agriculture. These practices must and will become the future of agriculture. Regenerative organic systems and Biodynamic agriculture have the power to reverse the current conditions and destruction of soils that have been done to date. All the destructive industrial farming practices will become extinct. It is just a matter of time.

The present industrial farming system with its broken nutrient cycle is anti-life, anti-nature driven practices that are not sustainable. It will inevitably collapse. It will fail. Only the planet's natural "soil food web" processes—involving bacteria, fungi, healthy plants, and numerous soil organisms that produce healthy fertile humus soil and plants—can feed our world. When fertility materials, minerals, carbon, and nitrogen are no longer imported onto the farm, health begins manifesting and a balance homeostasis begins to be restored.

So, I say again: we know what the solutions must be. To all our agricultural and health issues, we have conscious, best practice solutions that are capable, in one or two generations, of reversing environmental damage, regenerating resources, and restoring nature—our Mother Earth—and her soils. So, if we know what the problems are, and we have the solutions, humanity is dependent upon our living planet Earth for survival. Why is change happening so slowly?

Not only is the pursuit of global well-being misaligned with an ever-growing human population and increasing use of material resources needs. More importantly, the collective consciousness of humankind has also not grasped the degree of our dilemma nor the solutions we must undertake. We could say that consciousness is the only thing worth our consideration. Will the collective consciousness of the human psyche and society become sufficiently conscious in time? Will we act upon the environment for the future good of the individual and the whole? Every human must realize and participate in solutions to our current crises, or we are the problem.

As a species and as conscious, sentient beings, we were not always coming up this short. Today, we face a collective lack of education, poverty, nationalistic forces, corporate greed, and pollution of all kinds. Yet our current worldview falls far behind indigenous wisdom and Native Science, and we have a lot to learn from older cultures. We have still a lot to learn from nature. How can we create a better relationship with the living earth? Do we have sufficient individual and collective will to act upon known solutions for the good of the whole?

We must learn from our past. That is, we must recall the heritage that got us this far—that allowed humanity to reach our present state of development before ultramodern societal ills began to poison the proverbial well.

A Data-Based Reality Check

Where we are now as a planet, and where we are headed if we don't rapidly adopt available solutions, should alarm you. Decades of devastating deforestation rates should alarm you. Permafrost melting should alarm you. Soil carbon losses should absolutely alarm you. And we have the data to back all of this up. Two-thirds of the increase in atmospheric carbon has come from these three devastating phenomena.

Healthy forests enhance and support a healthy ecosystem and good agriculture.

For over one hundred years, in almost all managed agricultural soils—18% of the earth's land surface—we have seen rapid, ongoing carbon losses escalated by abusive human farming activity. Carbon is needed to support soil life, bacteria, and fungi that in turn feed higher organisms like protozoa and plants. Soil stability depends on sequestering carbon, carbon farming practices, and resultingly, in reducing atmospheric carbon. All of this is in very real jeopardy on a global scale.

The problems don't stop there. Oceans are overfished and polluted with nitrates, phosphates, radiation leakage, plastics, and manmade chemicals. Warming ocean waters have killed half of the coral reefs (bleaching) already, and we only have thirty years left before the other half is lost too. Many species are endangered—at least half of all the organisms on the planet are at risk of extinction, and the true number could be as high as 75% of all common species.

In short, the sixth mass extinction—the Anthropocene, human-driven extinction—has begun. If unabated, this is predicted to continue over the next few decades. This is the state of our global condition. *It is up to all of us to become well-informed, educated, and fast.* Will we allow our planet as we know it to become unrecognizable? Will we embrace life, and the forces of life? Or will we continue the unconscious path of death and global destruction? Our collective future depends on our current collective consciousness with the choices we make together now.

A Historical Foundation

You're probably asking: how did we get here in such a relatively short time? Historical context is critical.

From the beginning, agriculture was sacred knowledge. It was grounded in practices fathered by Zoroaster (Zarathustra) and taught by sages of ancient cultures as an art and sacred wisdom. The earliest story of agriculture tells us that agricultural wisdom was given to mankind by the goddess Nidaba (Nissaba) of Sumer, and in Greece by the goddess Demeter.

Agriculture allowed civilization to develop. Put another way, *without* the development of the culture of "Agri," our modern world and way of life would never have been made possible. Historical records show the earliest agricultural system existed in the Near East, known as Asia Minor in the Fertile Crescent, at the end of the Neolithic era and the hunter-gatherer culture. Subsequent systems of agriculture soon followed in Southeast Asia, the Nile Valley, and the Americas, fostering the growth of domesticated human populations. Will we preserve, and nurture the culture of Agri that has brought civilization this far on our journey of evolution?

Conscious Agriculture: A Solution

Conscious agriculture understands nature the need to align with nature and its influences as essential support for diverse agricultural practices—often involving smaller-scale agriculture, ecological, and biologically driven agriculture. Biodynamics, layout the original holistic ecological farming principles in 1924. The biodynamic method is unique because it addresses and embraces both a spiritual and scientific understanding of life and nature—much like the worldviews of global indigenous populations. In particular, the Celts and early Indo-Europeans. This way of thinking and growing is our inheritance—humanity's legacy.

Biological agriculture is *process agriculture* utilizing an unbroken nutrient recycling system as opposed to non-renewable Chemistry-based *input agriculture.*

The Australian Demeter Bio-dynamic method was primarily developed by Alex de Podlinsky in the early 1960s. In his lecture *Biodynamics Agriculture of the Future,* he lays out a holistic approach to biological agriculture, its balanced vision of agriculture of the future being healthy in harmony with Nature and our planet's health. Here we have advanced methods to regenerate soil and nature using special homeopathic-like materials.

Many other biological agriculture initiatives exist as well. The Rodale Institute's long-term organic vs. conventional trials, known as the Farming Systems Trial (FST), present a low-till, organic cover cropping method driven by building, maintaining, and revitalizing soil fertility. It has proven, for example, that we can produce all necessary corn and soy without any of the industrial-chemical practices presently decimating the environment and dominating destructive agriculture in America. Of the three systems studied conventional, legume cover cropping and the organic manuring system was the most profitable farming system, even without receiving premiums for organic crops produced.

The solutions expand from there. Regenerative organic or eco-agriculture, carbon farming, agroecology, permaculture/restoration agriculture, keyline design, agroforestry systems, alley cropping, silvopasture, well-managed livestock grazing, edible food forests, and smart farming techniques all have an important role to play in rethinking the monoculture system of industrial agriculture. Holistic management of these practices restores grasslands and the environment while building productivity through planned livestock grazing, land planning, and ecological monitoring. Together, these systems make up the growing regenerative agricultural alternatives widely in use today. Learning from nature and seeing nature and agriculture as sacred are the perspectives needed to change our materialistic destructive ways of growing food.

To be clear, we are in exciting times. All hope is not lost. More and more groups and stakeholders are rallying, leading the movement of conscious ecological agriculture (or agroecology), and calling for these practical, proven real-world solutions and conscious alternative systems of agriculture to be put into wide practice. A network of researchers and grassroots biological farmers, foodies, chefs, the Slow Food movement, and conscious consumers are demanding healthful quality locally grown food. Better quality, biological-organic, and/or biodynamic food is in demand—the growth of both organic and biodynamic food sales is expected to double within just five years, exceeding the extraordinary growth of the last three decades. In short, public awareness and demand for clean food will be the turning point to a new nature-based or nature-centered agricultural global model. We are learning from nature. This is conscious agriculture evolving!

Conscious agriculture is essentially about human consciousness —our evolution both individually and collectively as a species. Conscious farmers who respond to our looming concerns around food security, nutrition, farm worker justice, and environmental dilemmas are leading the conscious ecological agriculture way. We must catalyze our efforts to see change and progress by halting and reversing climate change.

At present, it is understood that soils sequester about thirty percent of global carbon emissions. With better, conscious soil-building strategies, the carbon burden in the atmosphere and the oceans can be reduced. Our solutions must be numerous and spiritually inspired, not driven by the materialistic thinking that has brought us to our current crossroads.

Our spiritual environmental consciousness will also be demonstrated by individual acts. Actions can be taken up by people who are driven and passionate about creating a better world: one no longer threatened by a global climate catastrophe; one without hunger; one that is in harmony with our natural environment. The ways we eat, purchase, and grow our food daily can have a great environmental impact. Learning from, cooperating with, and aligning with nature are the only ways to a conscious food future. We have the choice to create a world responsible for good stewardship of our soils and care for the environment, by acting consciously. Is it just the planet we need to save? Perhaps, is it humanity that we must be saved from our self-destruction?

As Franklin D. Roosevelt put it, *The nation that destroys its soil destroys itself.*

Practicing conscious agriculture encompasses many alternative strategies. While global adaptation is a tall order—no one disputes this—it is our only path to saving our world, a world full of potential and promise. Perhaps what this all comes down to is embracing eco-spirituality. It is an illusion that we are separate from nature. Embracing fully conscious agriculture is embracing nature and biological living soil, and taking responsibility for making big choices and moves required to halt and reverse the extreme climate changes that threaten human civilizations' co-existence with nature. Agriculture fueled by wisdom, not greed, should be our legacy.

Our fate is in our own hands. How will we shape our world for a better outcome? The future of food is either a macabre-dead, inhumane phenomenon driving us toward mass destruction, or it is a bright and thriving food system in harmony with nature and life. Our vision of the future of agriculture will determine if our food comes from green pastures, healthy fields, gardens, or materialistic food factories of death.

The question is often asked: why do farmers farm, given their economic adversities on top of the many frustrations and difficulties normal to the job today? And always the answer is: "Love. They must do it for love." Farmers farm for the love of farming. They love to watch and nurture the growth of plants. They love to live in the presence of animals. They love to work outdoors. They love the weather —maybe even when it is making them miserable. They love to live where they work and work where they live. If the scale of their farming is small enough, they like to work in the company of their children, with the help of their children. They love the measure of independence that farm life can still provide."

We cannot allow the global scale of industrial agricultural destruction to be the future of farming. The future of food. The future of humanity and our shared planet. Not when we have the solutions at our fingertips, waiting to be put into broad practice. We don't have any time to waste—let's dive in.

The conscious farmer considers their local community, the finances of their business, and the impact on those who will farm the land after they have gone. Conscious agriculture and farming are holistic practices involving long-term planning. Regenerating farmland, soils, and pastures by working with nature is at the core of conscious agriculture, but this does not and cannot happen overnight. So, the conscious farmer will consider the impact of every decision that they make. They think holistically about the soils and water on their farms and the wider environment. They care for their cows and livestock. They farm with joy and dedication. And, the bottom line is this: conscious farmers don't just know how to sustain the world's soils; they create and regenerate them.

Carbon Farming

Carbon farming is a common practice in ecological farming because carbon is a major, essential component of healthy soils. Carbon is the main food for plants (atmospheric carbon), carbon as dead plant material mainly feeds soil life, bacteria, fungi, and earthworms. In the form of dead plant materials, and plant exudates carbon feeds soil life, soil then stores more moisture and nutrients, supporting better soil structure. Organic carbon materials contribute to changes in soil structure (aggregates) and porosity helping increase rainfall infiltration. Carbon feeds the soil food web and the web of life. We speak of a carbon life form, meaning all life forms on earth. Carbon is the central building block for all biological molecules like proteins, DNA, and carbohydrates; essentially, every living thing we know is considered a carbon life form. Carbon is the element behind all forms with calcium as mineralized carbon the second great builder of form. Silica Si, as being the other great element behind forms on the earth.

Globally, enormous amounts of carbon have been lost to the atmosphere from excessively cultivated soils. Fossil fuels have since the year 2000 added, over 36 billion tones, of carbon to the atmosphere, with each year more being released, in 2024 over 38 billion tones have been added. A tipping point and warning is that adding another 200 billion tons in the next years will be a limit after leading to further irreversible weather devastation. But the good news is that carbon can be put back into the soil where it belongs, which is ultimately good for the climate.

One promising conscious strategy? Biochar, or charcoal. Historically known as Terra Preta, a black, fertile Amazonian soil created by indigenous Indians—the best of human ingenuity. The productive soil's black color comes from weathered charcoal and broken pottery, bones, compost, manures, and fire ash. Biochar is a natural carbon material with small pores that hold water, nutrients, and communities of beneficial soil microbes. It is very stable and slow to decompose, capable of binding minerals and soil nutrients with soil organisms for perhaps thousands of years, creating stable, porous soil. It is a soil aggregate, including small rocks and/or plant material. It can be useful when applied to poor soil environments (i.e. sandy soil, heavy clay, and rain-leached tropical soils).

Biochar is made by burning wood, and, admittedly, making it is an imperfect process. Specifically, wood materials must be burned slowly and 50% of the carbon is unavoidably lost to the atmosphere in the process. Then, in most cases, the burned wood must be hauled to its intended site, a labor-intensive process. Then, the process required to incorporate biochar into native soil requires great soil disruption or cultivation. Is it practical to take these steps, and at what costs? On a global scale, perhaps not. But in some cases, it is an amazing solution. While incorporating biochar into soils on an industrial scale would be challenging and very costly, it is ideal in some soils (sandy, or heavy clay) and wonderful in raised beds or smaller garden plots.

Animals, Livestock, and Land Restoration

A self-contained farm is a beautiful aspect of conscious agriculture. While soil is the foundation of a healthy, conscious farm and garden, true conscious farming is also about creating healthy pastures that the livestock feed from—which in turn has positive effects on human health when we choose to eat pasture-fed animal products. Well-managed layered livestock farming is becoming common as part of regenerative agriculture. Polyface Farms, in Virginia, managed by Joel Salatin and his family is a teaching and working farm utilizing layered livestock production.

All eco-conscious farmers need to appreciate and support the three key animals of fertility, according to biodynamics and Dr. Rudolph Steiner. It is essential to incorporate these three animals as integral participants in all eco-farming systems: the cow, the domesticated honeybee, and the earthworm. More than all other creatures, these three animals support the fertility of agriculture inside and out. The cow is a cloven-hoofed ruminant *upon* the ground. The earthworm is the great digester of mineral soil and organic matter *below* the ground. The honeybee is the greatest pollinator *above* the ground. Each of these animals plays a vital role in the farm's fertility and healthy productive agriculture.

Nature has its own inherent design for feeding the soil food web and the soil biome: organic materials and plant residue from production crops, legumes/cover crops, the grazing of pastures, and livestock manure all return soils to abundant living systems. In addition, healthy plants play an important role in promoting healthy soil by the release of exudates. To heal our climate and our planet, we need to mimic nature's design in several strategic ways. The good news?

Progress is underway. A process called land restoration is taking place on more than forty million acres globally, demonstrating the efficacy of using livestock to restore grasslands and reverse desertification.

Holistic management and planned grazing—a combined strategy taught and practiced by Alan Savory and the Savory Institute—is an essential land restoration practice that should be implemented globally. By restoring soil moisture, water catchments, and river flow, this system helps to increase forage, livestock, and wildlife populations. It utilizes low-stress animal handling techniques, restoring damaged and degraded land, and improving crop yields through concentrated animal impact.

Cover-Crop Planting

Organic management which includes cover cropping, improves soil structure and tilth, increases water infiltration, and does not contribute to the accumulation of toxins in waterways. Eliminating toxic pesticides by use of diverse cover-crop planting, for weed management will also fuel soil biology supplying carbon and nitrogen—thus creating healthy biologically active soil. Planting legume crops (pulses, such as lentils) in rotations can remediate up to 180-200 pounds of nitrogen per acre. Planting pastures with rotations of clovers and alfalfa eliminates the need for chemical nitrogen applications, altogether. These systems have long been known and used in sound, practical agriculture.

The wide use of biodynamic preparations applied to well-managed properties will bring profound changes to soil life, plant vitality, livestock health, and the produce that comes from well-managed biodynamic farms. At a very low cost, extremely small amounts of natural materials applied in homeopathic quantities can do it! Our goal should be restoring and maintaining soil harmony, fertility, and soil health while raising nutritious food.

Minerals, Soil, and Nutrition

The best small to medium-scale regenerative organic growers—especially biodynamic growers—cooperate with the biological forces of Nature and encourage the essential life-cycling of nutrients. They encourage the biological activity of the soil, developing soil aggregates, vigorous plant root systems, and greater plant root exudates that will slowly bring essential mineral nutrients back into the living sphere. Minerals are essential for optimum plant health and for a plant's ability to resist diseases and insects. Both livestock and human health need and depend upon mineral-dense plants to provide us with good nutrition.

Plants require a minimum of thirteen to eighteen (13-18 perhaps up to 42) essential minerals for healthy growth and development: These are phosphorus (P), potassium (K), calcium (Ca), magnesium (Mg), sulfur (S), iron (Fe), manganese (Mn), zinc (Zn), copper (Cu), molybdenum (Mo), chlorine (Cl), nickel (Ni), and boron (B) with sodium (Na). Some food plants like celery, beets, spinach, artichokes, carrot, broccoli, strawberry guava, goji berries, passion fruit, pineapple, figs, and dandelion for example, are good sources of Na sodium and potassium. The minerals P, K, Mg, Ca, Fe, Mn, Zn, S, Cu, Mo, Co, Cl, Ni, and B with Na are therefore the base to grow healthy plants and nutrition. A total of eighteen elements including oxygen (O), nitrogen (N), carbon (C), and hydrogen (H) are involved. When a plant has a full spectrum of essential minerals available from healthy soil the fruits and forage from these plants will be mineral-dense foods. Beneficial soil microbes need healthy plants as much as plants need to be healthy to support beneficial soil organisms. It is worth repeating again, that "more life is better", this is the true web of life!

The conscious farmer understands specifically: that plants are powered by the sun. Sunlight regulates the plant's sugar production influencing the plant's ability to take in adequate soil minerals. Plants must get their minerals from active living soils in which the minerals come in the ion form through natural biological processes. Plants take in micronutrients with fine white feeder root systems and are regulated by sunlight. Earthworms, soil bacteria, and fungi are responsible for soil carbon digestion, they release minerals from rock particles for plants to use. (Side note: Earthworm activity and their castings produce the cations that are present in the soil, via fungi and bacteria in the digestive tract of the earthworm). Both cations and anions are required for plant utilization—that is, plant roots absorb the soil's colloidal minerals when minerals are in super fine ionic form. Our human cell membranes operate along the same principles as plant cell membranes. Minerals within our bodies need to be in an ion form to pass through cell membranes and to carry out functions in our physiology at a cellular level.

Along these lines, the average consumer is being deceived by unconscious growers relying on rock power as inputs. The bio-conscious farmer understands that minerals were intended, by nature, to be food-sourced following nature's biological systems. But when the unsuspecting consumer buys mineral supplements as rock powder—in non-ionic form—they don't realize they are receiving very little value. A 1-2% absorption rate is common for most mineral (rock-sourced) supplements. From a 1,000 mg calcium supplement, our bodies likely absorb only 10-20 mg at best. In contrast, we can absorb 300-400 mg from a food-sourced 1,000 mg of calcium from food—a 30-40% absorption rate, from broccoli, for example.

Why does this all make such a difference? It goes back to considerations of life versus death. A plant's intact life processes allow it to absorb super fine colloidal ionic minerals. Colloidal minerals enable the plant to reach its full health potential, and thus provide nutrients needed to develop quality human food and good livestock forage. In other words, for plants to be healthy and thrive, a continual supply of super-fine minerals must be available in ion form each growing season. Biological farmers know and respond to this truth. When poisonous pesticides are used on crops, they kill the life processes that operate within the soil and plants. In addition, conventionally grown crops that are fed water-soluble nitrogen do not develop fine white hair-like feeder roots and do not absorb essential micronutrients from the soil. So, when the compromised plant cannot feed itself, conventional agriculture applies chemical, water-soluble fertilizers to *feed the plant*. Plants fed via this unnatural feeding system develop a tap root process for water uptake of nitrogen (in a toxic form), in which they absorb little or no micronutrients and, thus, nutrition is compromised. It must be noted that food grown locally and via conscious methods of eco-agriculture are also unprocessed foods. Ultra-processed foods created by industrial food processing, are foods deficient in essential minerals. These foodstuffs should be avoided altogether whenever possible. They are foods with conditioners, additives, and colorings, which are unnecessary. Highly processed foods are designed for consumers to eat more than the normal need of consumption of calories, such as soft drinks, many unnatural juices, convenience, snack foods, and ready-to-eat packaged and frozen meals.

Minerals and Health

One of the keys to health, longevity, and disease-free living is the ingestion of minerals from foods that are wild, or grown via biological organic, and biodynamic methods. Mineral-rich, nutrient-dense foods must be properly grown on mineral-rich living soils, these are high-quality foods. Conversely, the mineral-deficient plant and person will live a shorter life plagued by a myriad of symptoms and illnesses (dis-ease). It thus stands to reason that having an informed understanding and appreciation for the minerals that are most important to the healthy functioning of the human body's fluids, organs, and tissues will assist us in making more advantageous nutritional consumer choices. People with a mineral-dense diet will have longer lives, good health, better fertility, and greater vitality. Throughout the whole of our lifetime, minerals are continually being lost from the body, such as potassium and magnesium found in all fruits and vegetables, and must be replenished, primarily through nutrient-dense food. Minerals must come to us from our foods—supplements derived from powdered rock can never be a suitable substitute. The path to nutrition begins with sunlight and healthy soils abundant in soil biology. When all the minerals are available to the plant in living soils, healthy plants rich in minerals will be the result. Thus, human and livestock health along with seed health become possible.

We receive our first nutrition and minerals while in our mother's womb. The fetus in the womb is fed from their mother's blood supply, and directly after birth, breast milk should supply our nutrition. So, while pregnant and in the postpartum period, the dietary choices a mother makes will impact the fetus/child—influencing the health of the child for their whole lifetime. The mother's good eating habits provide the foundation and earliest building blocks for the child's health and longevity.

Children—especially infants—must receive the best biological-organic and biodynamically grown foods to build a good health foundation for life. For infants, the absence of a mother's milk replaced by a *scientifically formulated* replacement formula is never desirable. (Goat milk is even much richer in minerals than human milk and some believe it is the next best practical choice if a mother's milk is unavailable for a baby. Some nurses and pediatricians disagree, of course, with being schooled in a non-holistic science.) For school-aged children, nationwide, we have (perhaps unsurprisingly) dropped the ball on good food choices and optimum nutrition. Fortunately, the demand for better nutrition in public schools— quality, nutritional food options and the elimination of junk and fast foods in public lunchrooms—is being driven by conscious parents and activists. Unfortunately, corporate food marketers have children often desiring highly processed, sugary, low-quality nutritional food options.

Let's talk about potassium and phosphorous. Potassium "K" (Kalium)— with chemically twice as much alkaline action as sodium. Potassium—is known as the great alkalizer. This important element is more essential than ever in modern Western diets. Potassium is necessary to balance the high intake of "P" phosphorous in protein-rich food most Americans and Westerners eat. (P phosphorous, the predominant acid mineral in our diets, will never be found in any supplement form, often consumed in excess by modern people who are eating excess quantities of protein from diets of meats, seafood, eggs, and dairy foods. In addition, most soft drinks contain liquid phosphoric acid. The kidneys need potassium most, as K is lost daily buffering acidic phosphorus and uric acid from our systems. To take in sufficient amounts of potassium, government dietary guidelines advocate eating a minimum of five portions of fruits and vegetables daily—most fresh plant foods contain very little or no phosphorus and are good sources of potassium. The next best source of potassium is found in concentrated plant foods such as dried and fresh apricots, beans, soy, nuts, grains, seeds, and avocados. By and large, we have a long way to go in the UK and America, the average consumption of fruit and vegetables is about three and a half portions per day. This is a recipe for kidney disease and eventually kidney dialysis. The result of a lifetime of potassium deficiency (contributing to high blood pressure and cardiovascular diseases), and poor kidney health, is crippling our healthcare system. Note, that today one in five Americans dies from kidney disease (kidney failure). A diet of potassium-rich foods is key to good health. A diet of nutrient-dense foods grown locally on healthy soils will give the best advantage for overall well-being and longevity.

We must also avoid infective shortcuts. Drinking alkaline water is a very expensive and ineffective way to address our body's alkalinity —it only scratches the surface. At best, we absorb 1% of our minerals through the water we drink. Plus, many alkaline "mineral waters" contain aluminum, and, bicarbonates, alkaline elements that are neither desirable nor needed by our systems. Why waste money buying expensive alkaline water? Carefully, read the labels of the food products you purchase—alkaline waters are another example of the consumer being misled.

Calcium— Ca, or lime, as it is known in agriculture—has received a great deal of attention in nutrition and health care. Unfortunately, calcium is greatly misunderstood in modern society. While quality calcium *from food* is essential, we absorb calcium poorly from milk and dairy. And not only is milk falsely promoted as a good source of calcium, but it is also high in P phosphorus. Milk and dairy can easily be eaten in excess because milk and dairy products contain one part Ca to every two parts P (1:2 ratio) when we need a 2:1 ratio of Ca to P to absorb calcium efficiently. (While our bodies need equal amounts of Ca and P in the system, calcium has only a 30% absorption rate and P phosphorus has a 70% absorption rate. Yogurt is the best dairy food, with an absorption ratio of 1:1.) It's all a very delicate balance. We must reduce overall phosphorous in our diet to absorb calcium efficiently. Excess dairy or milk can increase P phosphorous in an already high-P diet, disrupting calcium absorption.

Consider some real-life health implications here. Because high phosphorus can lead to the loss of calcium, osteoporosis—a loss of bone density—is caused, in part, by the highly protein-rich acid diet of many Western consumers. The common loss of magnesium and calcium is directly due to a lack of plant-based, K-potassium-rich foods in a person's diet. Good calcium absorption and the prevention of its loss can be assisted when K potassium, Mg magnesium, and trace amounts of silica are abundant in our food choices and diets.

What are some steps the average consumer can take now? Eat a diet rich in alkaline or base foods, more fruits, and vegetables. More of a plant-based diet can be advantageous for many people and the environment. Support your good health and do your shopping at the local farmers market, local farms, and community food coops to obtain fresh biological-organic and biodynamic produce whenever possible. Choose to buy from regenerative organic and biodynamic farmers, who understand that K potassium is widespread and plentiful in nature—it is found in all plants (fruits and vegetables). When we recycle our food waste, leaves, branches, and kitchen scraps to make compost, we are recycling K or potassium. There is no need to purchase potassium for your soil when we use good compost, which is abundant in recycled plant material rich in carbon and potassium.

Data Guiding Us Forward

The Farming Systems Trial (FST, Rodale Institute) is America's longest-running, side-by-side comparison of organic and chemical agriculture. Insights from the trail are rapidly becoming an important part of a conscious eco-farming strategy.

After more than forty years of side-by-side research, Rodale Institute's FST has demonstrated that organic farming is better equipped than conventional agriculture to feed the world's population—both now and in the future. A myth has been perpetuated that organic, conscious agriculture cannot be a large-scale solution because it cannot produce food at the scale that conventional agriculture achieves. But FST data consistently demonstrates that after an initial decline in yields during the first few years of organic transition—which is typical in dead soils that have been managed chemically for decades—the biological-organic system soon rebounds to match *or surpass* the yields of the conventional, chemical-based system. Another myth exists that organic agriculture is a costly system. FST data shows that toxic nitrogen fertilizers underpinning conventional agriculture cost, on average, between sixty and seventy dollars per acre or more annually and are always increasing. Inputs purchased by conventional farmers decrease their net profits and their business viability.

If this kind of data can help convert more believers, we must disseminate it broadly. The stakes are quite high. For a century now, developed nations have followed the path of a death culture. While we used to be mostly unaware of the dire consequences bearing upon us. Yet, we are still wandering this precarious path at the beginning of the 21st century. The time is overdue that this march of disease and death stops cold. This is important: to restore soil and plant health, to restore our environment; it is also to restore, on a higher level, our nutrition, and support a balanced climate. As we face uncertain and extreme weather patterns, increasingly expensive diesel fuel supplies, a growing global water shortage, and ever-expanding human populations, we will require farming systems that can not only keep up but also adapt, regenerate, and transform the land. That is, we need nature's systems that can not only withstand new, mounting environmental pressures but also mitigate them while producing healthy, nourishing food.

Can food that is grown in a conscious agricultural system be the food that might also help raise human consciousness? Conscious agriculture is responsible agriculture with the ability to respond! That is, respond to the need for local food independence to provide much of a community's food and nutrition. One potential solution to disseminate the (literal and figurative!) fruits of conscious agriculture throughout communities across the country, and the world, is the establishment of educational farms and hands-on training farms. Food Arks, or Ark Farms, inspired by the Agrarian Resource Collaborative (A.R.C.), the ARC farming project. An Ark farm is a self-sustaining and productive "farm individuality" with many diverse plant and animal species—a working farm system that can feed its community and provide vital local food. Existing such farms include Turtle Tree Farm, Seed Savers Exchange, Threefold Community Farm, in New York, and new projects like the Scattered Seed Project, which are valuable examples of Seed Arks. The biodynamic model might be the ideal Ark farm, the system that introduced Community Supported Agriculture now four decades ago, CSAs operating as self-contained farms are stable ecosystems feeding local communities. Conscious agricultural systems can simultaneously strive for the three-pronged goals of education, collaboration, and economic stability. Ark farm properties and CSAs should be held in land trusts, as foundations, and educational non-profit local farms. The farms of the future must be "seed saving" or seed preserving farms, being sources for both animal seed and plant seed preservation, thus making them farms for humanity.

In her book *Agroecology and Regenerative Agriculture: Sustainable Solutions for Hunger, Poverty, and Climate Change*, Dr. Vandana Shivas synthesizes decades of agroecology implementation research. Dr. Shiva states that agroecology is an effective solution not just to climate change, but also to many other ecological crises humanity faces, such as water security, land degradation, seed saving, and biodiversity loss. At present, about 50% of all seeds are controlled or "owned" by a few global agrochemical-pharmaceutical corporations, and addressing the ownership or control of seeds is a fundamental right for all of humanity.

Ecological farms and biodynamic farms, provide habitat for wild or feral bees important pollinators. On biodynamic, regenerative-organic farms, home backyards, and gardens the planting of grass pastures, interplanted with flowering legumes, flowering hedgerows, windbreaks, filter strips, crops in rotation, and more. It all comes down to this: biodiversity is the pathway back to nature's bounty and, ultimately, to our nutrition. Nutrient cycling of elements is essential to balanced soil, plant, livestock, and human health. Aligning with nature, the farmer and gardener work to renew and cycle essential elements (NPK) and other essential minerals (Ca, Mg, Si, Mo, Fe, Mn, Zn, Co, B, and S).

Our Shared, Bold Vision

Our individual, actions and our voices have the power. When we grow *or* purchase healthy biologically grown organic and biodynamic produce, we take important steps to support conscious agriculture. Together aligning and working with nature, we can stop and turn around the climate crises, all while feeding the world nutritious food grown from ecological farming systems. With this strategy, we can have a brighter future.

> *The most important thing is to make the benefits of our agricultural preparations (biodynamic preparations) available to the largest possible areas over the entire earth, so that the earth may be healed, and the nutritive quality of its produce improved in every respect. That should be our first objective.* (E. Pfeiffer, 1958, p.120)

While this important, bold vision is yet to manifest on all cultivated global land, Steiner's Agricultural Course presented 100 years ago, can nevertheless already be deemed a success. Steiner's ideas have been transformed into international organic and biodynamic movements within sixteen years or less from their 1924 unveiling, and his ideas remain at the innovative, leading edge of burgeoning regenerative organic, ecological farming, carbon farming, and agroecology movements. Dr. Steiner's revelations have survived, evolved, developed into practical methods, and proliferated. Their influence persists intact as the worldwide biodynamic Demeter movement.

We must have a clear and concise working knowledge of the biosphere (we are working with many good theories) that can lead us toward real solutions to global ecosystems' health and harmony. The time has come to evolve more consciously in our agriculture practices on a grand scale, to transform our outdated industrial-chemical agricultural mono-crop farming systems while demonstrating ecological stewardship.

THE POWER OF NATURE AND THE FUTURE OF AGRICULTURE

BIODYNAMIC, REGENERATIVE ORGANIC AGRICULTURE AND HOLISTIC MANAGEMENT

So, what is the answer for agriculture? What is the great and promising path forward to our planet's redemption? How can we generate a return to harmony between humankind and nature?

For agriculture to be successful, those of us participating in its practices must know and cooperate with the ways in, which nature works. We must align with Nature! Put more directly: conscious agriculture holds the keys we must implement to collaborate with nature.

"Nature can provide for everyone's need but not for greed" Mahatma Gandhi

Let's take into consideration these facts. The average conventional farm can produce two thousand or three thousand dollars gross per acre per year with a net profit of perhaps $450 to $250 if growing corn or soy on better years. Farm earnings can be far less. In 2023, the average farmer return was projected to be as low as $25 per acre for these same crops. Depending on the year and if higher or lower prices are paid for their commodity crops. An organic farmer can generate twice as much in income per farmed acre. The organic farmer produced 10 to 18% less yield, yet their incomes were much higher. "MN-WI (Minnesota and Wisconsin) organic row crop producers managed 325 acres per farm on average and generated $324.24 per acre in median net farm income in 2020 and 2021." "The 2020-2021 data are very promising for small-to-mid-size farms," says Joleen Hadrich, professor in the Department of Applied Economics at the University of Minnesota (UMN). Conventional row crop farms, by comparison, were more than three times as large; averaging 1,053 acres per farm while netting $144.65 per acre in median farm income."

At UC Davis, ranked number one in the US in agriculture and forestry, we have a small experimental sustainable-organic farm. Their CSA program is reported to generate as much as $17,500 annually, per acre in produce. Many smaller organic CSA farms nationwide can generate twenty to thirty thousand dollars per acre in produce or more. One of the highest-grossing production farms that I have found is Elliot Coleman's biological-organic "Four Season Farm", a commercial market garden and farm in Harborside, Maine. Farm production has generated $220,000 on his two acres netting, thirty-five thousand dollars annually per acre. And number one in productivity is "Singing Frogs Farm" in Sebastopol, California. Producing year-around and generating (grossing) $160,000 per acre per year, on an organic intensive production no-till/low-till farm. This production is possible with a super-efficient cultivation system and very high soil fertility by incorporating generous amounts of local organic compost. The farmers, Elizabeth & Paul Kaiser, report soil fertility as high as 11% organic matter in their vegetable beds. This may seem unprecedented, but 11% organic matter was achieved at Alex Podolinsky's farm in Powelltown, Victoria, Australia, using biodynamic preparations and his farm compost. Alex was unwilling to publish these soil fertility findings to avoid criticism and the aggravation of dealing with skeptical agricultural academics, that he found to be closed-minded. Alex stated, that after thirty years of biodynamic farming success and soil regeneration, no academic or state agricultural employees ever showed interest in the ADBM successes.

The Rodale Institute's Farming Systems Trial (FST), 40-year

report gives us affirming studies for supporting organic-biological cultivation and soil building. "The results clearly and consistently demonstrate that organic management protects and builds the health of soil. The health of crops, the environment, and farmers are positively impacted, while conventional practices inevitably lead to degradation of the soil and diminishing returns for farmers." "The scientific data gathered from this research has established that organic management matches or outperforms conventional agriculture in ways that benefit farmers and lays a strong foundation for designing and refining agricultural systems that can improve the health of people and the planet." This ongoing research shows that organic practices increase soil organic matter, microbial biomass, diversity, and activity while reducing soil compaction. It further shows organic yields match conventional yields for cash crops, such as corn and soybean. This debunks the myth that organic agriculture can not compete with conventional systems. The data proves that organic system operation costs are significantly lower than conventional management and certified crops receive a premium return.

Embracing Nature's Power

Inherent to the conscious, biodynamic philosophy is a recognition that rather than attempt to change nature, we must work *with* nature. No matter what we wish to do, nature will always find its own way— so our acceptance of its needs is necessary before anything else can begin.

Agriculture has an important role to play in creating harmony with nature. This is the bottom line: we can no longer continue to contribute to nature's destruction. Instead, we must do the opposite. We must mindfully and actively work toward nature's restoration— all the while never underestimating nature's power. We have all the solutions. We have the solutions for a secure, stable future for the planet and humanity. What are those transformations? It includes a rapid transition away from fossil fuels. It is a transition towards circular business models, it is a change to creating healthy nutritious diets to organic and biodynamic food systems. We must halt the losses of Nature and scale the regeneration of marine systems, our agricultural soils, forests, peat bogs, and wetlands.

Nature is inherently vital, resilient, regenerative, competitive, transformative, and at times, destructive. When nature's environments are given a chance—that is, when human activities cease, are rolled back, or are managed biologically in accordance with nature—soil and all plants and animals will respond and rebound, becoming bountiful and returning to their natural state. Understanding how ecosystems are connected is an essential component of this process. We must reconnect and communicate with nature and all its connected beings.

The thread of love that connects us to nature and all life, can and should be our legacy, as a species. Biodynamic farmers, producers, and consumers alike will find that a love for nature and the living world functions as a powerful intention that can change and enhance our relationship with agriculture. In short, we must treat nature and all living things with respect and dignity—and this includes the planet itself. As is often the answer, consider ancient wisdom. The Earth is seen as a living being by many indigenous cultures and some insightful conscious farmers, ecologists, and biodynamic practitioners.

The future of agriculture lies in aligning with nature and ceasing our attempts to dominate it with one-sided farming of monocrops, and CAFO operations that stress and deplete the environment. Conscious farmers' proper management and handling of domestic livestock and farm manures, has the ability, to create and maintain soil health. Mimicking nature's wild livestock herds of cloven-hoofed animals, cattle, and buffalo with well-managed grazing is essential. Nature will respond by rebuilding habitats, biodiversity, and environmental vitality.

Going Forward = (Rightfully) Going Back

Our lives depend on the bounty that agriculture can provide, and it is the farmer or agriculturalist who primarily determines the nature of such activity today. In ancient times, this was an exalted role. The person who held the knowledge of *life's mysteries* or processes—those who could provide food and sustenance by successfully raising livestock, and growing food crops—was respected and revered. They were seen as priest-like for their role in society. But today, the same cannot be said of the modern population of aging, lowly farmers or farm workers who are, for the most part, looked down upon as lower-class citizens or simply essential physical laborers. Unfortunately, they are to be replaced by more machines and AI in the not-so-distant future.

In English, the root word (Anglo-French "fermier" (masculine) "fermière" (feminine)) for *farmer* means *"to pay rent" as a tenant*. In the origin days of our feudal fief system, in a still-used agricultural system today, the farmer paid for and *rented* the land for use from the lords—the landlords and landowners, often royalty or their vassals. This was the fief system, just a step away from slavery. Today, farmers in the industrial system face similarly demeaning conditions as a minuscule cog in a global machine. But in contrast, the local eco-organic grower and biodynamic farmer is held in high esteem. He or she is revered as the knowledgeable steward of life and its laws, capable of and responsible for feeding your family. The conscious farmer has achieved, within this oldest of professions, a fulfilling and highly respected role in society again.

Wise agriculture and conscious farming will be the only method of growing food in the future. There is no other option. Humanity can no longer afford continued ignorance and greed in growing food, at the expense of nature and the environment. We need to make rational conscious choices available to us now.

Lest it seem that ecological farmers are backward practitioners, in reality, conscious farmers are true trailblazers. As part of their daily operations, they use both today's innovations—passive solar energy, innovative wind energy technology, and bio-diesel fuels—as well as proven, traditional practices. They understand that agriculture must return to the use of draft animals, smaller farm machinery, and appropriate farm equipment designed by eco-farmers and ecologists, rather than industrialist engineers. They understand that animal production must be conducted on a smaller, manageable, healthy scale. Plant and animal materials must be returned to the farm's soil.

Industrial-mono agriculture in all its forms must be rejected and repudiated. Environmental businesswoman Maria Rodale's 2010 *Organic Manifesto: How Organic Farming Can Heal Our Planet, Feed the World, and Keep Us Safe* is well worth reading. It underscores and calls to action many of these same belief systems: that conscious, moral, principled individuals must champion stopping the degradation of nature in the process of growing our food; that we must neither participate in nor purchase industrial chemically produced foods any longer; that regenerative organic farming, ideally leading to widespread regenerative-organic and biodynamic agriculture, must be our choice and our shared goal; that our food choices must reflect conscious participation in restoring the land and our planet.

Animal Husbandry the Path Forward: Holistic Management

The principles of Holistic Management and planned grazing, originally advocated by Alan Savory, demonstrate a fundamental and actionable basis for good animal husbandry. Unlike some other conscious practices that must still be rapidly scaled, this holistic management system is already widely introduced and thriving, having been implemented on over fifteen million hectares worldwide. (This begs the question: why does the system still lack scientific validation from academics?) While there are always critics, skeptics, and naysayers, many practical farmers are already benefitting from applied practices of planned grazing and layered livestock pastured grazing. This train has, thankfully, already left the station, no matter what industrial systems try to do to stop it!

In Mexico's Sierra Madre ranching country, this system is being practiced on large cattle operations benefiting from planned grazing. In the great southwestern region of the United States, the system Holistic Resource Management (HRM) was an official agricultural management system *for the state of Utah.* Joel Salatin of Polyface Farm in Virginia teaches and has implemented a viable layered livestock operation, grazing dense herds of cattle, hogs, and poultry.

While animal rights advocates, vegans, and vegetarians are a relatively small percentage of our modern societies (with India being the cultural exception), animal rights advocates have powerfully expressed justified environmental and health concerns regarding the treatment of domesticated animals by, industrial agricultural practices. Confined Animal Feedlot Operations (CAFOs) have greatly violated the mandate of good animal husbandry. In contrast, many people desire and thrive on pasture-raised beef and meats. At times I have referred to animals as "people"—or, even, as the best of people —as they are our kin, whether we admit it or not.

As a lifelong vegetarian-pescatarian myself, and as a farmer who was raised on a cattle ranch in Montana, I respect the wisdom of animal husbandry. Cattle are integral to good agriculture—nature's inherent ecology utilizes cloven-hoofed animals to critically benefit its ecosystems. A gross modern, materialistic error has been making meat production the primary motive behind rearing livestock; we will all benefit when this is no longer our systemic way. Today, we currently use half of all grain produced globally, on half of all agricultural soils growing grains, to feed animals bred for meat production in CAFOs. *This is a non-sustainable system.*

For the livestock farmer, grass-fed beef, goats, sheep, hogs, and poultry are economic products. But in the industrial kind of system, animals are not allowed to tend to the land in ways they naturally would. Overgrazing or under-grazing creates mismanagement of pasture lands, and the long-term result of poor land management is desertification (drought) which detrimentally alters the climate. Improper grazing of livestock has contributed for thousands of years to desertification in the Middle East, North Africa, the Mediterranean, and now America's great southwest. Livestock should be raised on pastures both for their health and for (primarily) the benefit of the environment. And when this happens—when animals are raised for their warmth, companionship, manure, and the many dairy food products they provide to humanity; when they are raised naturally and not pushed for maximum food production—they exist and function in harmony with nature. All ranged bovine or herding-based cloven-hoofed operations can benefit the environment when managed using a conscious planned system.

The very first agricultural practice was the herding of livestock for two thousand years. Most human cultures have their origins in deep connections and meaningful relationships with the creatures of nature. The cow is sacred in India—and rightly so! Native Americans revere the buffalo not just for the meat they provide, but for the whole animal that sacrifices its life and provides hide for leather, hairy skins for blankets, bones for tools, and shelter. Eastern cultures and indigenous peoples speak about a sacred force that is present everywhere and in everything, believing that nature and all things are sacred. Nature services are many, giving to humanity and the planet's ecosystems in numerous ways. We do not need to return to the ways of ancient hunter-foragers or indigenous cultures and their ways of life entirely. However, the wisdom of indigenous people needs to be understood and respected in our materialistic times. We need to have sympathy for and a conscious understanding of the deeper values that lie behind the culture of the hunter-forager and the agricultural traditions of animal husbandry.

A Data-Based Reality Check

Where we are now as a planet, and where we are headed if we don't rapidly adopt available solutions, should alarm you. Decades of devastating deforestation rates should alarm you. Permafrost melting should alarm you. Soil carbon losses should absolutely alarm you. And we have the data to back all of this up. Two-thirds of the increase in atmospheric carbon has come from these three devastating phenomena.

Healthy forests enhance and support a healthy ecosystem and good agriculture. But our forests are being logged and burned relentlessly, destroying valuable natural ecosystems around the world. Already, some forests are only one percent of their original glory. Five thousand years ago it is estimated that 75% of England was forested and today it is at a new low of just 10% forest remains. And our standards have plummeted: when 25% of an original old-growth forest remains, that's now considered substantial. One-half of the earth's forests—or 2.6 trillion trees—have been harvested or lost. When human beings developed civilization, it is estimated that there were 5.1 trillion trees on the planet.

Peat bogs are amazing carbon sinks—they store ten times more carbon dioxide than any other ecosystem. Earth's peat bogs, mostly found in the temperate northern hemisphere, cover 3% of the planet's land surface and hold enormous amounts of sequestered carbon (66%). As permafrost melts and recedes in northern latitudes, the largest global reserves of carbon and methane gases are being released into the atmosphere. In addition, peat deposits are being mined globally on a mega-scale, and this is very concerning. Mangroves globally hold 3-5 times more carbon than forests, yet they have been lost or threatened and diminishing on all coasts. A few restoration projects are attempting to halt their destruction further.

For over one hundred years, in almost all managed and cultivated agricultural soils—18% of the earth's land surface—we have seen rapid, ongoing carbon losses caused by and escalated by abusive human farming activity. Carbon is needed to support soil life, bacteria, and fungi that in turn feed higher organisms like protozoa, beneficial nematodes, anthropoids, earthworms, and plants. Soil stability depends on sequestering carbon and, resultingly, in reducing atmospheric carbon. All of this is in very real jeopardy on a global scale. Growing and nurturing healthy plants and as many diversified plant species as possible is a major key to sequestering carbon in soils on farms and all landscapes.

The challenges don't stop there. Oceans are overfished and polluted with nitrates, radiation leakage, plastics, and manmade chemicals. Warming ocean waters and nitric oxide (from atmospheric acid fallout), have killed half of the coral reefs already, and we only have thirty years left before the other half is lost too. With a one to two-degree average global warming as the new normal more environmental losses are expected. Many species are endangered—at least half of all the organisms on the planet are at risk of extinction, and the true number could be as high as 75% of all common species.

The sixth mass extinction—the Anthropocene, human-driven extinction—has begun. If unabated, this is predicted to continue with irreversible consequences over the next few decades. This is the state of our global condition. *It is up to all of us to become well-informed educated, and fast.* Will we allow our planet as we know it to become unrecognizable? Will we embrace life, and the forces of life? Or will we continue the unconscious path of death and global destruction of life forms? Our collective future depends on our current collective consciousness, and on the choices, we make together now.

A Historical Foundation

You're probably asking: how did we get here in such a relatively short time? Historical context is critical.

Agriculture allowed civilization to develop. Put another way, *without* the development of the culture of "Agri," our modern cities, world, and way of life would never have been made possible. Historical records show the earliest agricultural system existed in the Near East, known as Asia Minor in the Fertile Crescent, at the end of the Neolithic era and the hunter-gatherer culture. Subsequent systems of agriculture soon followed in Southeast Asia, the Nile Valley, and the Americas, fostering the growth of domesticated human populations.

The earliest form of agriculture was animal husbandry—shepherding of domestic livestock. Good animal husbandry—was foundational to the first waves of subsistence agriculture systems. Livestock dung or manure, and well-managed grazing practices provided the basis for the soil's increased fertility; when livestock herds were properly managed, they resupplied and recycled the essential plant nutrients necessary for agriculture to thrive.

For a long time, traditional agriculture was always a natural, "organic" practice. It didn't involve any synthetic materials (substances derived from Petrochemicals and concentrated nitrogen compounds) and little if any artificially imported fertility materials. In short, agriculture succeeded for centuries practicing simple, minimalist, or low-till soil cultivation, mostly recycling and regenerating plant nutrients.

And then came the Industrial Revolution, which ushered in practices of industrial agriculture—and everything changed.

Just over two hundred years ago, along with widescale societal changes caused by the Industrial Revolution, we began burning exponentially more and more fossil fuels—coal, petroleum, and natural gas. At the same time, the advent of "chemistry-based agriculture" began the accelerated practice of death in soils and our environment. Since the Industrial Revolution began, excess atmospheric carbon has doubled. *At the present levels of consumption of fossil fuels in thirty years atmospheric carbon may double again!* Are we listening?

These trends of decline accelerated exponentially from there, in line with major world events. In the 1950s and 60s, after WWII, growers began using water-soluble nitrogen fertilizer on a grand scale. Toxic poisonous chemical sprays (insecticides, fungicides, and herbicides) followed, a relatively new threat to life. Today, millions of tons of deadly chemical cocktails are used annually in the production of most of the world's food supplies completely unnecessary. It is estimated that the average American was eating with their foods conservatively one pound of these synthetic chemicals annually, according to the Environmental Working Group (EWG), ewg.org a non-profit environmental organization since 1993. And as much as 2.5 pounds of these man-made synthetic chemicals annually (announced 40 years ago by two research doctors, (a biologist, and a medical doctor) that won a two-million-dollar grant for demonstrating the detoxification of cancer-causing PCBs from human fat tissue. In contrast, the human body in a lifetime produces less than one ounce of hormones from our glands. Imagine the biological nightmare the human body (your DNA and chromosomes) are subjected to, with ingesting these chemicals found in your common foods purchased at the supermarket!

When will these extremely deadly chemical practices cease? And how will Nature and the Earth recover? Our ancestors relied one hundred percent on sunlight as our fuel and food source. Today half or fifty percent of our proteins and much more of our calories from carbohydrates rely on fossil fuels to grow, produce, and sustain modern humans! We must become environmentally conscious very soon as a species to reverse the destructive trends we humans have unleashed.

The Future of Agriculture

Small to medium-scale farms are the future of agriculture. There is no way that farming equipment can cultivate a farm of three thousand acres, or more, without compromising fertility, soil structure, and the environment. It is the wrong scale. The system that has adopted these large-scale operations has been designed to eliminate or minimize human labor at great cost to nature. This is the unhealthy scale of agriculture common today.

In good farm management, timing is everything. If a farmer must till the earth from sunup to sundown on every possible day available, soil erosion, soil compaction, and degradation will inevitably occur.

Conscious farming on biodynamic farms is appropriately scaled and balanced with natural production with the integration of crops, pastures, and orchards. Put another way, for a conscious farmer to make a living wage, he or she grows crops to achieve a sustainable and manageable scale. However, there are cases where biodynamic production capacity can be *appropriately* scaled, where the activation and application of biodynamic preparations have been scaled up using stirring machines and very efficient spraying equipment. For example, biodynamic regenerative sheep and cattle ranches producing grass-fed livestock can operate using biodynamic preparations on many thousands of acres as demonstrated in Australia.

The food we grow and eat must nurture our bodies—our physical temples—but how can we expect such nurturing when such a vast majority of today's available food is linked to a culture of death and destruction of life? It should be self-evident that these two paths are inherently contradictory. We can't be sustained, nor can food provide health for our bodies, while our systems and practices support the destructive, life-diminishing agriculture that dominates our present culture. The highest values held by conscious, free people—our health freedoms, and our spiritual freedoms—must be met and reflected in how we treat and interact with nature and all its creations. Producing nutritious food has been the exclusive goal of biodynamic agriculture—its main tenant since biodynamics' inception one hundred years ago—so there is no denying that biodynamics is the way forward to a more conscious agriculture.

Society, and agricultural practitioners, cannot be free as long as we are codependents of a toxic, deadly addiction to synthetic nitrogen compounds (fertilizers), petroleum, and its petrochemical byproducts that contradict everything that nature and life are about. An addict must first acknowledge that they have a problem or addiction, to initiate the measures that will eventually lead to freedom from their addiction. Taking actions that free us from our toxic dependency is neither easy nor convenient, but it is intelligent, conscious, and essential that we implement these practices now. We must show our respect for the complex myriads of creations all around us. We must give our respect to nature and all its life forms. This must be our legacy.

Biodynamic agriculture, as presented by the late Alex de Podolinsky of Australia, has addressed the global dependency of "organic agriculture" on chemical (albeit non-synthetic) inputs. More than forty years ago, Podolinsky reminded leading organic farmers at an IFOAM international conference, that it is their responsibility to strive to create humus soils produced by biological activity—that is, to create living soil, the only real future for agriculture.

The Nitty Gritty: Details We Must Understand

Plants are beings of the sun and earth and thus obey the laws of nature. In all its interconnected, symbiotic processes, it is obvious that nature has an intelligent ordered connectedness and relationships at work. We may, perhaps, call this universal consciousness.

The power of nature will be found in the details. They say the devil is in the details, but we—the movement of biodynamics and conscious individuals—know it is actually, love found in the details.

Would you believe that plants are the most intelligent beings on earth? Yes, decades of scientific studies have shown that plants are the most intelligent beings on earth. Plants are miracle workers, the great chemists of nature, and have created many symbiotic relationships with soil microbes. Plant roots deliver many complex chemicals and nutrients to soil microbes via "plant root exudates". The more plants, and the more diverse plant species the more soil microbes are fed, this exponentially multiplies and proliferates the soil fertility.

As we know, the earthworm is one of nature's great co-workers—the MVPs of soil fertility, if you will. In large part, their immense impact is due to the contributions they make to soil cations. Cations are present in the soil largely because of the highly productive fungi and bacterial activity in earthworms' digestive tract. They are the minerals Mg, Ca, Fe, K, Zn, and Mo—in the electrically charged form plant roots require to absorb essential alkaline minerals from the soil.

To drive home so much of what we have covered across these chapters, let's roll up our sleeves on this culminating topic. Soils carry an electrical charge, and good soils have a greater charge than poor soils. Plants benefit enormously from this soil electrical phenomena.

Cation exchange capacity, or CEC, measures the soil's ability to hold cations by electrical attraction. Cations are positively charged elements, the positive charge indicated by a + sign after the element symbol. (The number of + signs designating cations indicates the amount of charge the element possesses. For example, for our purpose, relative to the five most abundant exchangeable cations in the soil: calcium (Ca^{++}), magnesium (Mg^{++}), potassium (K^+), sodium (Na^+), zinc (Zn^{2+}), and aluminum (Al^{+++}).

Cations are held by negatively charged particles of clay and humus colloids. Colloids in silica clays consist of thin, flat platelets. They have, for their size, a comparatively large surface area, making them capable of holding enormous quantities of cations. They act as a storehouse of nutrients for plant roots. As plant roots take up cations, other cations in the soil water replace them on the colloid. For example, if there is a concentration of one cation in the soil water, those cations will force other cations off the colloid and take their place. The stronger the colloid's negative charge, the greater its capacity to hold and exchange cations, hence the term *cation exchange capacity (CEC)*.

Clay is an essential element in healthy well-balanced soil. Clay particles or platelets are made up of many layers of colloidal silica crystal sheets that have an amazing cation-holding capacity. Clay has a great capacity to attract and hold cations because of its chemical structure. CEC varies according to the type of clay: it is highest in montmorillonite clay, found in chocolate soils and black puggy alluvia; it is lowest in heavily weathered kaolinite clay, found in krasnozem soils, and slightly higher in the less weathered illite clay. Low CEC values can be improved by adding organic matter (compost/humus).

(On the other hand, Sand has little or no capacity to exchange cations because it has no electrical charge. This means sandy soils such as podzolic topsoil have very low CEC, but this can be improved by adding organic matter (humus compost and or humic acids) or good compost. This is an example of how an eco-conscious farmer works *with*, never against, the power of nature.)

Soil pH is important for CEC because, as pH increases (becomes less acidic-more alkaline), the number of negative charges on the colloids increases, thereby increasing CEC. Humus CEC levels vary according to the type of soil. Humus is the end-product of decomposed organic matter, waste residue, and metabolites from plants, animals, and microbes, and organic matter colloids have large quantities of negative charges. *Humus has a CEC 2-5 times greater than mont-morillonite clay and up to 30 times greater than kaolinite clay, so humus colloids are very important in improving soil fertility.* The lesson here is that soil improves when organic matter, compost, and humus are built up in all soils.

Calcium or lime is often applied to acid soils to balance acidity and stimulate or promote plant growth. Many species of earthworms produce millimeter-sized aggregates of calcite crystals called granules. These calcium crystals are alkaline and help to balance the soil's pH so that it does not become too acidic. Earthworms prefer and thrive in a more neutral pH 7 soil, or a little more alkaline. Plants require the element calcium to grow, and many soils have little or no calcium, so the earthworms are organizing a better, healthier soil for themselves and plants benefiting the Earth. How earthworms, manifest calcium in the soil—building up this mineral element when there is no detectable calcium, to begin with in many soils—is still a mystery. Materialistic science has not and cannot provide an answer for this phenomenon. Dr. Steiner may have given clues as he stated calcium is *mineralized carbon* and was the first element to manifest the Earth's early crust, precipitating from our earliest gaseous planet's atmosphere.

Good, conscious farmers, agronomists, and scientists recognize the role that nitrogen—the "N" in NPK—plays in so many aspects of agriculture. Nitrogen is plentiful in nature. (It makes up 78% of our air or atmosphere.) More broadly, nitrogen is easily available in a balanced, biologically driven farm in an active and stabilized form in living soils (when biology processes are actively at work). Soils rich in carbon feed soil aerobic bacteria—this is the path to "N" fertility.

In a closed farming system, leguminous plants (pulses) are one of the primary sources of nitrogen fixation. Other sources providing soil organisms with carbon and nitrogen include animal manures and composted or mulched green plant residue. Along with properly managed cultivation of a high number of earthworms, nitrogen-rich soil makes for good, fertile soil capable of supporting the successful, commercial-scale growing of virtually any agricultural food or fiber crop. Whenever possible, high-quality, compost (stabilized nitrogen) produced on the farm using biodynamic herbal preparations should be made and regularly applied.

All Comes Back to the Soil

Decades-long compost trials conducted by the California State University Davis (CSUD) School of Agricultural demonstrated the greatest soil-building tool for a farmer was regular applications of good organic compost. Understanding why compost works—and why chemicals that attempt to replicate or replace it do not—is one of the most important concerns of conscious agriculture. When we consciously choose and refuse chemical fertilization, we instead rely on the soil food web's own complete and inherent interactions to provide crops with the nutrients they require for proper healthy, and sustained growth. Once again, *harnessing the power of nature* is the fundamental and successful premise of biological-conscious agriculture.

Compost provides stabilized nitrogen in humus compounds (humic and fulvic acids)—that is, food for the soil food web. High-quality organic compost—but *especially* biodynamic compost—also inoculates the soil with beneficial microbes, bacteria, and fungi (with complex long-term food sources for soil microbes). Biodynamic compost provides subtle, critical influences that become *available* to all the important minerals over time. Many incredibly rich interrelationships and secrets of manuring (as Steiner referred to fertilizing) are now being discovered not only by soil microbiology researchers such as Dr. Elaine Ingham, soil microbiologist but also by working-conscious farmers and growers. As we know, nitrogen is at the top of the nutrient list for plant growth. Namely, how and in what form plants receive nitrogen from the soil is of vital importance in agriculture. But nitrogen requires help from other essential minerals to do its essential work. The nitrogen from living organisms is the building block of amino acids. As the transport system for getting amino acid nitrogen into the plant tissues, silica plays a crucial role, through fungal hyphae. The element calcium is also needed, to help activate nitrogen. Boron and sulfur trace elements allow for calcium to call nitrogen and move it into the root zone through the silica-rich tubes of the mycorrhizae.

Ultimately, the natural web is beautifully complex. It makes rational sense that chemical fertilizer *substitutes* can never compare. These chemical fertilizers destroy the nitrogen-fixing microbes and the fungi that help move the nitrogen in the soil. The affected plant will take up nitrate, wasting the plant's energy and quickly needing more nitrogen. It should become apparent why conventional farmers buy so much nitrogen: once they've been misguided by the *experts* to use chemical fertilizer, they must keep using it much as a drug-addicted person must keep using the drug they have become addicted to. What else happens when large quantities of chemical nitrogen (nitrates and ammonia compounds), when added to soil, making it necessary to use chemical nitrogen again and again? Nitrogen-fixing microbes are burnt and killed. Fungal growth is suppressed because nitrate is toxic in excess quantities. Photosynthesis is inhibited, delayed, or blocked. And of course, the chemical-promoting experts' *salesmen* are paid through ag-chemical-drug companies for promoting this unnatural system of chemistry agriculture.

The biodynamic preparations, compost, and compost teas all come to the rescue and inoculate soils with the beneficial microbes that agrochemicals killed. In addition, cover cropping, crop rotations, and livestock-animal manures, quickly revive farmland and healthy crop growth. The soil food web is nature's system vital to the web of life. The conscious ecological farmer avoids chemical soil compaction (soil hard pans), death-promoting harmful chemicals, over-acidification of soils, and all else that harms soil life.

Let's take it one step further, to a fundamental principle offered in Steiner's agricultural lectures that has, received little attention. Plants should take in elements via humus and not through the soil water. Steiner meant that this process is driven by sunlight; and that how nitrogen reaches plants matters. *The more life the better* is another statement attributed to Steiner, meaning more biodiversity, thus more plants equal more roots, and more roots equal more plant root exudates with a mirid of nutrients that feed the soil food web, creating more soil biology and greater soil aggregates building soil fertility.

As nitrogen is constantly needed for properly regulated, healthy growth, plants take it in slowly over the course of the entire growing season. In water-soluble feeding of plants, the tap roots are inundated with high amounts of nitrogen during early growth and become deficient in or depleted of a stable nitrogen supply over the later stages of the growing season. Thus, a plant becomes watery and susceptible to a myriad of preventable diseases. On the other hand, healthy plants develop many fine white feeder roots and take up stabilized nitrogen from humus soils on a stable, ongoing basis. These truths are unknown by most conventional soil scientists and farmers. However biological farmers and growers must be intimately familiar with this fundamental knowledge of soil biology to become free from the deception of chemical plant cultivation.

As liquid or concentrated nitrogen is not in accordance, with nature's laws, it cannot harness nature's power. But in healthy soil, it is unnecessary. Humus soil contains sufficient nitrogen in various forms, existing as amino acids. In biologically healthy soil, nitrogen is supplied by decomposing organic matter, including dead bacteria and soil organisms, dead plant residue, protozoa excretions, dead wildlife, animal livestock manures, and earthworm castings. Nitrogen is released and dispersed when any higher soil organism excretes or dies, fixing in the soil via bacteria and living symbiotically on the roots of leguminous plants. Bacteria found active within the earthworm's digestive systems also supply plants with the nitrogen that they need.

Although nitrogen is challenging to secure in soils, it is plentiful in the environment in various forms. Water-soluble fertilizers shut it down; conscious farming practices bring it back. However, phosphorous—the P in NPK—presents similar challenges and is not so widely spread in nature. Petrochemicals also destroy potassium and phosphorus-fixing microbes that supply these nutrients to crops.

Wherever phosphorous (P) is required, material reductionist thinkers would never agree that anything but the physical importation of P to a property would suffice. But on some farms where P has been applied for decades, its accumulation or build-up has caused many to rethink its continual annual applications. Soil tests from farms long dependent on the annual application of P show that plant growth (without deficiencies) is achievable after not applying additional P for eight to ten years. If sustainable farming is practiced, only 25% of most annual P applications will be sufficient. Alex Podolinsky of the Australian Demeter Biodynamic Method provides a compelling example. He talks about a citrus (grove) farm that was converting to biodynamic practices after using conventional methods for 30 years. Soil analysis demonstrated that 97% of the phosphorous purchased on the property over three decades was still in the soil, but the plants on the property had used only 3% of all that had been applied over those three decades. Then, try extrapolating the millions of farms worldwide where similar conventional systems are practiced. What an ill-advised waste of money and resources is the conventional agriculture system! The monetary costs are passed on to the consumer. The physical costs are passed onto nature—and the great waste of a valuable element of nature is criminal and shameful.

Conscious Agriculture's Environmental Impact

It is widely recognized that compared to conventional agriculture, organic agriculture generally has a positive effect on a range of environmental factors, like biodiversity, increased soil carbon sequestration, soil quality and structure, soil erosion mitigation, better soil hydrology (less water runoff and reduced flooding), and the reduction of climate change and global warming contributors.

Even biology, its scientists, and biodynamic farmers have not yet penetrated all the real possibilities behind the power of nature. Yet, each year, we continue to learn more. Cover cropping and more biodiversity are very beneficial to a regenerative system of agriculture. Eco-farmers and biologists are implementing evolving insights with the newly revealed knowledge of the soil food web. Nature-based science and spiritual science may hold many answers, but knowledge and wisdom are still much needed by humankind. This is exactly why nature restoration is essential to eco-agriculture consciousness: the more we understand how nature's systems work, the more we can cooperate and even collaborate with them.

Charles Darwin, the noted scientist and father of the theory and science of evolution, said in his final book, *The Foundations of Vegetable Mould, Through the Actions of Earthworms* (1881): "It may be doubted if there are any other animals which have played such an important part in the history of the world as these lowly organized creatures."

The Role of the Consumer

Each of us always has choices to make regarding becoming stewards of conscious agriculture and cooperating with natural systems of agriculture. And today more than ever, conscious farming choices are many. It is now up to the individual to decide to be a conscious local and global citizen consumer: driving solutions, supporting nature refusing to continue participating in the proliferation of industrial-chemical problems with our food and health.

It will require a conscious society of individuals who fully embrace conscious agriculture for the needed changes to bear fruit. Environmental farming's differentiating ability to nurture biodiversity and restore the landscapes and environments *is the future*, for biodiversity is the fabric that holds our lives together. Humans are completely dependent on nature and its biodiversity. And while the growing number of conscious farmers and eco-minded growers are making a difference, the tide will still only be fully turned when great numbers of individual consumers jump on board ecology-based agriculture as well.

Australia: Our Model Going Forward

As we see in Australia—a global biodynamic leader—is the world leader in organic agriculture---conscious farming practices are making a real impact. Permaculture, organic, and biodynamic farmers have turned the tide of fortune for Australia. Once the most environmentally precarious continent, it has now become the greatest producer of conscious agriculture, certified organic foods, and certified Demeter biodynamic foods.

If the agricultural systems of Australia, with its poor soils and challenging environments, can provide *more than 50% certified organic-regenerative, biodynamic food grown worldwide,* then the other continents with unique but still challenging soils and environments have no excuse. The vast majority of them can and must rapidly become free of chemically driven systems of industrial agriculture. We know how soils can be regenerated and productive.

Our Mandate, and Our Hope

Is the beginning of the end of industrial chemically driven agriculture in sight? Dr. Vandana Shiva—Gandhian eco-activist, humanitarian, seed-saving advocate, and scientist, ---founder of Navdanya (meaning nine seeds)—an earth-centric, women-centric, farmer-led movement advocating for the protection of biological and cultural diversity and seed sovereignty as protection for human health. Their philosophy? That the web of life is a food web. Heirloom seed holds all the genetic diversity for the future and life. [i]Hybrid seeds lack up to half the nutrients or more of heirloom seed crops, thus nutrition is a major factor in why we choose regenerative-organic, biodynamic, and eco-farming systems as these farmers are perpetuating heirloom seed and crops.

---

[i] Moconony, Industry Scandal: The Loss of Nutrients | Full Documentaries

We are interconnected with natural systems and all living things—we survive and thrive by way of an ecological process of co-creation with other life forms. That's why both the planet and people deserve a poison-free world. The health of the planet and our health are one and the same. — Dr. Vandana Shiva

There is also a great deal of creative positive action and promise being undertaken by educators and artists with the Organization of Nature Evolutionaries (O.N.E.). They envision a future where people and nature are co-creative partners, and all life has the right to thrive. Where nature and humans can cooperate and evolve with each other, listening to and building relationships with the living earth.

We must discover wonder in the beauty and diversity found in nature. Wonder is a spiritual quality that we need to embrace—*when our view of nature is a sacrality, seen with reverence and wonder, we open our hearts and minds to new possibilities.* When agriculture is in harmony with nature's profound truths, then we cannot destroy nature. And it is surprising that, collectively, we know so much and utilize so little of our great knowledge! Moving forward into a redemptive future, the eco-conscious small farmer and gardener will become the guardian and assistant of nature and biodiversity. Put simply, the future of agriculture—and hence the future of our species and our environment—lies in the hands of conscious, proactive individuals acting for and in full cooperation with the power of nature! The future of a healthy thriving planet Earth with all its diverse life forms intact is now, quite literally, in our hands.

To succeed on these critical fronts, we must always consider nature's rights. On this front, resources abound. The Movement Rights Indigenous Environmental Network the Global Alliance for the Rights of Nature and the Wild Law Institute coedit the publication *Rights of Nature & Mother Earth: Rights-based Law for Systemic Change.* Two additional important books further detail the pressing issue of nature's rights: *Uprisings for the Earth*, by Osprey Oreille Lake, and *Wild Law* by Cormac Cullinan.

We find ourselves on a confounding precipice. Will our human consciousness evolve or awaken to the reality that all life is consciousness? Is there a unified consciousness within all living things—as it seems to be—that will soon be realized? And why has our culture stopped valuing nature's intelligence? Why is only human intelligence and human manipulation of nature the only valued perspective? Is it the immaturity of human beings or the arrogance of our false superiority? Can we find our way back to nature? Can we give wild nature space and respect? As so many of our wisest minds have so eloquently tried to teach us, if we want our civilization to survive, Mother Nature needs our conscious thoughtful care and attention now:

There is no difference in the interests of human beings and the interests of nature. –John D. Liu, Journalist, Ecosystem Ambassador (Commonland Foundation)

The way we see the world shapes the way we treat it. If a mountain is a deity, not a pile of ore; if a forest is a sacred grove, not timber; if other species are biological kin, not resources; or if the planet is our mother, not an opportunity – then we will treat each other with greater respect. This is the challenge, to look at the world from a different perspective. –David Suzuki

A time must come in earthly evolution when it will be impossible for one individual to enjoy things on the Earth at the expense of another. –Dr. Rudolph Steiner

Cherish the natural world because you are a part of it and you depend on it –Sir David Attenborough

Look deep into nature and then you'll understand everything better. Embrace all living creatures and the whole of nature. –Dr. Albert Einstein

Forty percent to half of the earth's biodiversity is found in its soil. It is through and with the power and wisdom of nature that we can create truly humane, conscious agriculture: coexisting with nature while growing and producing nutritious, wholesome food. Will we see the connection between humans, humus, humility, and humanity? To be sustained, to thrive, and to evolve, human populations require wisdom. And it's not just the big world powers—it's each of us. We all must make conscious and wise choices that are aligned with nature to realize a future of food must be in balance with nature.

Zoroaster's teachings reflected the importance of the abundance and fertility of our natural world: "Agriculture is one of the noblest of all employments because he who sows grain sows righteousness" (quoted in Williams Jackson 1906: 373-4). "One of the most joyous spots on Earth is the place where one of the faithful sow's grain and grass and fruit-bearing trees, or where he waters ground that is too dry and dries ground that is too wet" (from the Avesta, Vendidad 3:32 & 3:4). The teachings and practices of Zoroaster place the respect for and stewardship of the environment in a central position of human civilization.

Will we see the beauty in nature, reflect, and give thanks for the creation of the natural elements on which life depends? Practice respect for natural resources by recycling and campaigning to clean up the air and water. Why not revive practical, possible practices, such as planting a tree as a celebration of birth and death? Nature has the solutions—when we see, listen, and work with her.

The time is now. It was yesterday. Our challenges on climate change were articulated very well in 1985 by Carl Sagan when testifying before the US Congress, and it is now almost forty years with little meaningful progress made. Given that we have the solutions available to us, we have no excuse. We must act now to avoid further loss for the next generations.

As of 2023, world governments have agreed on plans to protect and implement biodiversity for land and oceans by 2030. It is a start, but it's not the complete answer because at least thirty percent of our earth must be remitted to Wild Nature. If we are to achieve a return to all that our beautiful blue planet has to offer—if we are to exist and thrive in harmony with her once more—such agendas must be embraced alongside the widespread practice of conscious agricultural systems restoring and regenerating nature's soils.

INDEX

Entries in italics refer to published books and movies

A

Z

#

CONTACT INFORMATION

Contact Us: consciousagriculture@gmail.com
https://consciousagriculture.org/ or
https://conscious-agriculture.com/
Facebook: Conscious Agriculture: Aligning With Nature
Consulting, Collaboration, and Public Speaking Appointments
Phone: (01) 928-515-4151